Bilinear Control Processes

With Applications to Engineering, Ecology, and Medicine

This is Volume 106 in
MATHEMATICS IN SCIENCE AND ENGINEERING
A series of monographs and textbooks
Edited by RICHARD BELLMAN, *University of Southern California*

The complete listing of books in this series is available from the Publisher upon request.

Bilinear Control Processes

With Applications to Engineering, Ecology, and Medicine

RONALD R. MOHLER
DEPARTMENT OF ELECTRICAL
AND COMPUTER ENGINEERING
OREGON STATE UNIVERSITY
CORVALLIS, OREGON

ACADEMIC PRESS New York and London 1973
A Subsidiary of Harcourt Brace Jovanovich, Publishers

ACADEMIC PRESS, INC.
111 Fifth Avenue, New York, New York 10003

United Kingdom Edition published by
ACADEMIC PRESS, INC. (LONDON) LTD.
24/28 Oval Road, London NW1

Library of Congress Cataloging in Publication Data

Mohler, Ronald R
Bilinear control processes.

(Mathematics in science and engineering,)
Includes bibliographical references.
1. Control theory. 2. Automatic control.
3. Biological control systems. I. Title.
II. Series. [DNLM: 1. Mathematics. QA564 M698b 1973]
QA402.3.M63 629.8′312 73-2072
ISBN 0-12-504140-3

AMS(MOS) 1970 Subject Classifications: 93C40, 93B15, 92A05, 49C05

PRINTED IN THE UNITED STATES OF AMERICA

To my parents—David and Elizabeth

CONTENTS

PREFACE

This book analyzes the control of a significant class of nonlinear systems which are coined bilinear. Here, bilinear systems are described by ordinary differential equations that are linear in state, linear in control, but not jointly linear in both.

Bilinear control processes are very common in man and his environment. Several such systems are analyzed, and it is apparent that bilinear mathematical models are so useful to model real-world dynamical behavior because of their variable structure.

Taking a lesson from nature, it is shown that bilinear control can be utilized to man's advantage. For example, it is shown in Chapters II and III that bilinear processes are generally more controllable and offer better performance than linear systems. Here, sufficient conditions for complete controllability, based on the work of Lee and Markus, are derived. By means of the Bellman–Hamilton–Jacobi equation and the work of Boltyanskii, sufficient conditions are presented for complete controllability and optimal regulation of bilinear systems with bounded control. In Chapter III, a design procedure is suggested for practical suboptimal control, and it is applied to a tracking problem.

Optimal switching control is analyzed in Chapter III and shown to be particularly significant for systems that are linear in control. To compute optimal trajectories for such systems, which include the bilinear class, a recursive gradient alogithm is developed. The ensuing sequential solution is shown to converge to a maximum-principle solution. The advantages of this so-called switching-time-variation method are studied and several examples are considered.

Another computational method, called quasi-linear programming, is developed in Chapter III to solve nonlinear control problems. The method has proven effective for continuous optimal control computations for systems linear in control. The method is applied to a nuclear-reactor control problem, and the availability of canned linear programming routines (e.g., MPS 360 with IBM 360 computers) makes this alogorithm particularly useful for processes with numerous state constraints.

Besides bilinear nuclear fission and heat transfer studies in Chapter IV, several biochemical, physiological, and ecological variable strucures, which can be modified by bilinear systems of equations, are analyzed in Chapter V. Among others, these include cellular processes with enzymes producing bilinear control, transfer between organs with variable diffusivity of membranes causing bilinear control, temperature regulation of the human body with vasomotor control of circulation producing bilinear control, and regulation of carbon dioxide in the lungs. A brief discussion of socioeconomic systems and a cursory case study of the system prevailing in mainland China is presented in Chapter VI. Again, a variable structure such as offered by the bilinear system is most important for modeling, in this case for socioeconomics.

Chapters I–III establish a theoretical and computational base for bilinear control systems. This material is of particular interest to system scientists and engineers. Readers with a particular interest and training in an area of application may wish to skip certain sections in these first three chapters as suits their needs. Sections 1.4, 2.4–2.7, 3.3.2–3.3.4, 3.4, and 3.5 could be omitted for this purpose.

As a reference book or graduate text, the book should be of value to system scientists, engineers, ecologists, mathematicians, biologists, and physiologists. Much of the material has been presented in an advanced control course and in seminars in computer science, engineering, mathematics, and medicine. It is hoped that the book will open the door for more research and a better understanding of some very complex problems.

This volume has evolved from the author's NSF-supported research on bilinear systems, and he is grateful for their support. The author is indebted to his many associates and friends for their contributions to this work—in particular, R. E. Rink, S. F. Moon, H. J. Price, W. D. Smith, and R. S. Baheti. Also, the author wishes to thank Bev Dvorak for her patience in ably typing the manuscript, and his wife, Nancy, for her general understanding.

I

Introduction

The object of this book is to establish a need for a mathematical class of nonlinear systems that are termed bilinear and then to analyze the modeling and control of these systems. Bilinear control processes are shown to be quite common within man himself and throughout his environment.

In this chapter, a foundation is established for a bilinear system analysis of many problems pressing society today in such areas as medicine, ecology, socioeconomics, and engineering. Modeling and identification of systems are discussed along with a definition of bilinear systems and some of their relevant properties evolving from the Volterra series.

1.1 SYSTEM SCIENCE IN SOCIOTECHNOLOGY

The "systems approach" has been the coordinating force of multiple disciplines in harnessing the atom, in placing man on the moon, and in large-scale data processing. It is apparent that the same techniques are appropriate for many of today's ills which are of a more social nature.

For example, it is expected that computer-automated health-care centers connected in a nationwide grid eventually will reduce the cost of health care so that all Americans will have access to proper medical attention. Many corresponding large-scale system problems, such as the design of an optimal information system, are obvious immediately. Preliminary analysis for an important subsystem, automated patient records, is presented by Weed [1].

Even on a smaller scale, systems analysis is needed. Multiphasic screening may put incoming patients at a remote computer terminal to process background information and possibly make preliminary diagnosis or at least narrow down the possibilities. To make the preliminary diagnosis, new physiological data which are collected from tests on the patient may be integrated with the patient's record by means of standard statistical analysis, pattern recognition, and process identification. Physiomathematical models such as those derived in this book may be utilized with parameter norms for healthy patients. Such techniques will allow physicians to utilize their expertise more efficiently.

There are plans for a national biomedical communications network (BCN) which will couple the health-care centers. A center has been established in Lister Hill, New Jersey, by the Department of Health, Education and Welfare (HEW) to coordinate the development of this network. In the preliminary systems analysis it is necessary to compare the cost–effectiveness of common carrier BCN, dedicated terrestrial BCN, shared satellite BCN, and dedicated satellite BCN. The study requires mathematical modeling, identification, simulation, and optimization of system configurations. This problem is typical of many which involve multidisciplinary systems analyses. Patient and physician requirements, economic cost constraints, traffic density profiles for audio and video signals, reliability requirements, computer requirements, error and distortion control, signal processing, links to common carrier systems, and radio-frequency interference are among the parameters that need to be considered. Davis further reviews some of the requirements of a BCN [2]. It is urgent that expertise from the corresponding disciplines be integrated in a systematic manner to analyze such complex problems effectively.

Many large-scale systems analyses have fallen flat because they were too superficial and did not consider the real world. Far too often it has been convenient to base initial assumptions on mathematical convenience rather than on physical relevance. On the other hand, practitioners frequently have wasted funds and time on meaningless experiments and have made erroneous conclusions due to little or no understanding of system concepts. It is hoped that this book will not have either pitfall, and that it will help close this gap in some small way at least.

America's past Vietnam policy provides a good example of this common difficulty. As for so many politically oriented problems, policies have been obtained frequently by averaging individual solutions of numerous authorities. Historically, this technique goes back to the famous Oracle of Delphi, and is referred to in modern thinktanks as the Delphi method. In the case of Vietnam, the hawks and doves may both have had locally optimal solutions in their own right. Still, as frequently happens in mathematical optimization problems, an average of the two locally optimal policies may result in an extremely bad policy. For Vietnam, this apparently resulted in an ineffective middle-of-the-road policy. Such a dilemma is projected onto Fig. 1.1.

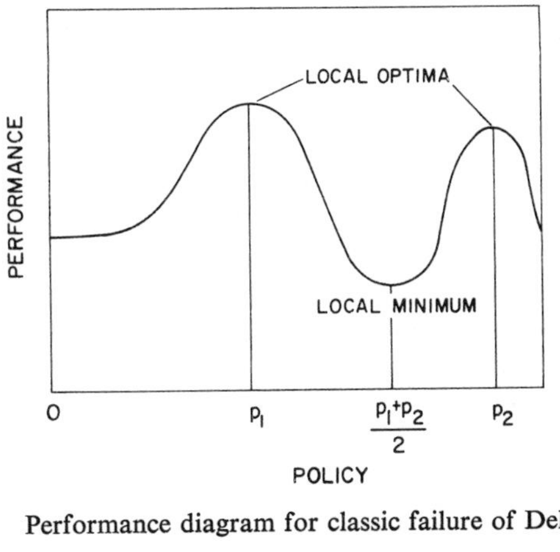

Fig. 1.1. Performance diagram for classic failure of Delphi method.

The role of systems analysis in urban processes is discussed by the Urban Engineering Study Committee of the American Society of Engineering Education [3] and in the area of control application by a NASA symposium [4]. It should be emphasized, however, that we are barely scratching the surface in these areas. The advent of the computer has added impetus to the many possibilities of relevant systems analyses.

Interconnected power systems and power plants themselves, such as the nuclear plants discussed in Chapter IV, are nonlinear in dynamics just as are those discussed above. Variable structures such as those studied here play an important role in the modeling of these processes.

1.2 MATHEMATICAL MODELING AND BILINEAR SYSTEMS

Effectiveness of any system study is only as good as the corresponding mathematical model. While such models may take a variety of forms, depending on the problem at hand, models considered here are described generally by ordinary differential equations or state equations of the form

$$d\mathbf{x}/dt = \mathbf{f}(\mathbf{x}, \mathbf{u}), \tag{1.1}$$

where $\mathbf{x} \in R^n$ is a state vector, $\mathbf{u} \in R^m$ is a control vector, the independent variable t is time, and $\mathbf{x}_0$ is a specified initial state for some initial time t_0 [5]. It is assumed that $\mathbf{u}(t)$ is at least measurable on an interval $[t_0, t_f]$, and normally that $\mathbf{u}(t)$ is piecewise continuous.

For the quite general nonlinear system described by (1.1), however, a detailed analysis presents a formidable task indeed. Some useful results have been derived with respect both to approximating solutions and to the stability of these systems. Some of the classical work is summarized by Coddington and Levinson [6] and Bellman [7]. More modern theories, based primarily on the work of Liapunov and Popov, are discussed in numerous publications including excellent texts by LaSalle and Lefschetz [8] and Aizerman and Gantmacher [9]. An application of this work is given in Chapter II, along with an analysis of controllability.

If the system is linear, as is so often assumed for convenience, (1.1) takes the form of

$$d\mathbf{x}/dt = \mathbf{A}\mathbf{x} + \mathbf{C}\mathbf{u}, \tag{1.2}$$

where $\mathbf{A}$ is an $n \times n$ constant matrix and $\mathbf{C}$ is an $n \times m$ constant matrix. It is natural that linear systems have been so thoroughly exploited since they possess well-known solutions to initial condition problems. Zadeh and Desoer [10] provide a good summary of linear systems analysis for the uninitiated reader.

Though linear models frequently are used to approximate real dynamical processes for convenience, they are inadequate in many cases. The simple equation which describes population of biological species is a good example. The rate of change of population is

$$dx/dt = ux, \tag{1.3}$$

where u, birth rate minus death rate, may be considered a control variable. For the so-called closed-loop population equation, the population coefficient u depends on the population or the state x, and (1.3) is nonlinear in state. Even for an analysis of the open-loop system, with u considered to be an input control, (1.3) cannot be put into the form of (1.2).

Similar to the biological population model, neutron population equations so familiar in nuclear fission processes take this bilinear form. Again, the equations arise in a natural manner by a simple economy balance which is studied in Chapter IV. As shown by Mohler and Shen [11, pp. 3–5], the neutron population n and neutron precursor (unstable fission product) populations $(c_1, \ldots, c_6)$ are approximated by

$$\frac{dn}{dt} = \frac{u(1-\beta) - 1}{l} n + \sum_{i=1}^{6} \lambda_i c_i \tag{1.4}$$

and

$$\frac{dc_i}{dt} = \frac{u\beta_i}{l} n - \lambda_i c_i, \qquad i = 1, \ldots, 6, \tag{1.5}$$

where u is the neutron multiplication factor, l is the average neutron generation time, β_i $(i = 1, \ldots, 6)$ is the portion of neutrons generated by the ith precursor group with $\sum_{i=1}^{6} \beta_i = \beta$, and λ_i is the decay constant of the ith group. (See Mohler and Shen [11, p. 4] for typical values.) Again, the system is linear in state, linear in control, but is not jointly linear in state and control as (1.2) is.

Since frictional force generated by rubbing surfaces is nearly proportional to the product of the orthogonal force between the surfaces and their relative velocity, the frictional force that is generated by a mechanical drum brake such as that shown in Fig. 1.2 is approximated commonly by

$$f_b = c_b u_1 \dot{x}, \tag{1.6}$$

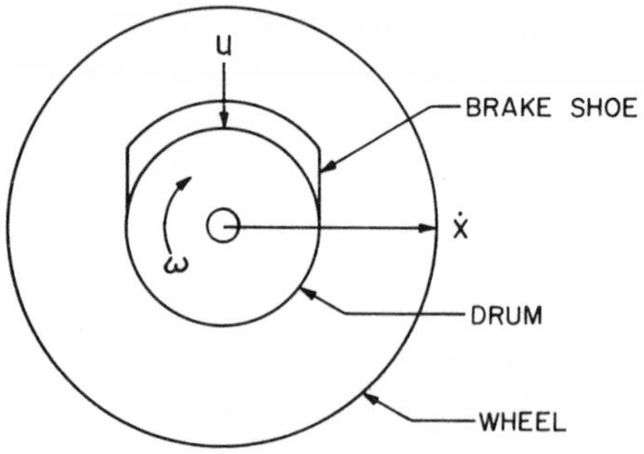

Fig. 1.2. Bilinear mechanical brake.

where the control u_1 is the normal force applied to the brake, $\dot{x}$ is translational velocity, and c_b is a constant. Here coulomb friction has been neglected. Similarly, other frictional forces may be approximated by

$$f_c = c_f \dot{x}, \tag{1.7}$$

where c_f is a constant.

By summation of inertial force, frictional forces, and engine force, the dynamical form of a braking car is given by

$$d\mathbf{x}/dt = \mathbf{A}\mathbf{x} + u_1 \mathbf{B}\mathbf{x} + \mathbf{c}u_2, \tag{1.8}$$

where

$$\mathbf{x} = \begin{bmatrix} x \\ \dot{x} \end{bmatrix}, \qquad \mathbf{A} = \begin{bmatrix} 0 & 1 \\ 0 & -kc_f/m \end{bmatrix},$$

$$\mathbf{B} = \begin{bmatrix} 0 & 0 \\ 0 & -kc_b/m \end{bmatrix}, \qquad \mathbf{c} = \begin{bmatrix} 0 \\ 1 \end{bmatrix},$$

k is a proportionality constant, m is automobile mass, c_f is a friction constant, and u_2 is the force due to the engine. Consequently, the driver controls the speed in the classic additive-control sense by adjusting the gas pedal, while a movement of the brake pedal results in parametric control of the system dynamics. It is shown in Chapter II that parametric control yields a variable-damping structure and that additive control allows a change in equilibrium states. Again, the system has the familiar bilinear structure. These systems are termed bilinear by the author and have the general form of

$$d\mathbf{x}/dt = \mathbf{A}\mathbf{x} + \sum_{k=1}^{m} \mathbf{B}_k u_k \mathbf{x} + \mathbf{C}\mathbf{u}, \tag{1.9}$$

where again $\mathbf{x}$ is an n-dimensional state vector, $\mathbf{u}$ is an m-dimensional control vector, $\mathbf{A}$ is an $n \times n$ constant matrix, $\mathbf{B}_k$ $(k = 1, \ldots, m)$ is an $n \times n$ constant matrix, and $\mathbf{C}$ is an $n \times m$ constant matrix. The state diagram of this system is given in Fig. 1.3. It is shown in the following chapters that these so-called bilinear systems are quite prevalent in nature, that they can be more controllable than linear systems, and that they can offer better performance than linear systems.

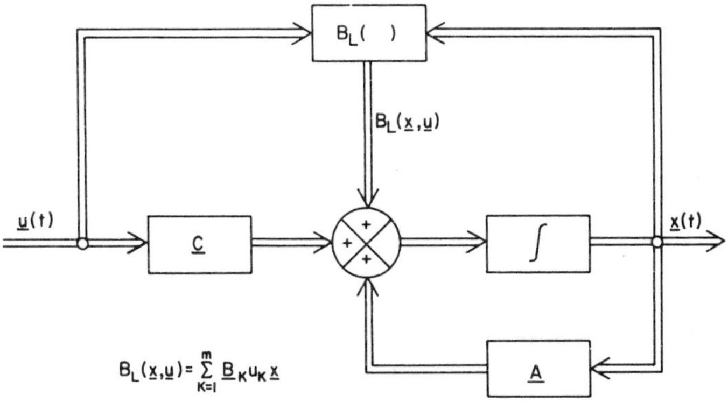

Fig. 1.3. Bilinear state diagram.

1.3 SYSTEM IDENTIFICATION

Bilinear modeling of relevant processes in engineering, medicine, and socioeconomics is studied in Chapters IV to VI. As for the processes considered in Section 1.2, it is shown that mathematical bilinear models frequently arise in a natural manner. And for more highly nonlinear systems, it provides another degree of refinement beyond that of a jointly linear model.

Even after the system structure or form of system equations is established, however, various parameters or coefficients of the differential equations must be estimated. For example, note the friction coefficients c_b, c_f in (1.8) or the mean generation time l in (1.4) and (1.5). Many parameters may be evaluated directly from published data which are normally available for precursor parameters λ_i, β_i in (1.4) and (1.5). Even these had to be obtained from experiments such as those presented by Keepin [12]. Others, such as l, however, must be estimated from data collected for particular system experiments [12]. Experimentation of this kind must take into account physical as well as mathematical conveniences and constraints. Some of these physical aspects are considered in Chapters IV to VI.

Current methods of parameter identification are often classified according to the mathematical techniques involved. Such popular classes of parameter identification for nonlinear systems are presented in Table 1.1. These techniques are appropriate for parameter identification of bilinear systems (1.9) and, of course, for linear systems (1.2) which are special cases of (1.9) with zero $\mathbf{B}_k$ matrices. Additional identification techniques are available for linear systems which utilize their nice analytical solutions in the form of transfer functions, impulse responses, or responses for other test signals. And indeed, the bilinear system (1.9) with a specified $u(t)$ is a linear but generally time-variant system. While certain linear system techniques may be appropriate to identify the linear structure or even the complete structure of some bilinear systems, such is not always the case. A word of caution should be noted here.

For example, consider a neutronic transfer function between perturbations in neutron population Δn and in multiplication Δu,

$$G(s) = \frac{n_0 \prod_{j=1}^{6} (s + \lambda_j)}{ls \prod_{j=1}^{6} (s + \rho_j)}, \tag{1.10}$$

where $n_0 = n - \Delta n$ is the equilibrium population for equilibrium multiplication $u_0 = u - \Delta u = 1$, ρ_j are the negative poles of $G(s)$, and s is a complex variable [11, p. 6]. Equation (1.10) represents the linearized system dynamics for the bilinear fission process (1.2). It is noted, however, that this system is structurally unstable, for any slight change in u_0 from 1 changes the characteristic behavior. In fact for any constant positive u the system transfer function has one positive real pole, and

TABLE 1.1

CLASSIFICATION OF NONLINEAR IDENTIFICATION TECHNIQUES[a]

Class	*A priori* structure required	Test signal dependent	Parameter constraints	Data processing
Coefficients in differential equation				
Differential approximation	Complete	No	Possible	Nonsequential
Parameter optimization	Complete	No	Possible	Nonsequential
Optimal control formulation	Complete	No	Possible	Both
Model reference	Complete	No	N/A[b]	Sequential
Functional series expansion				
Volterra expansion	None	Yes	N/A[b]	Nonsequential
Wiener theory	None	Yes	N/A[b]	Nonsequential
Analytical solution				
Parametric search	Complete	No	No	Nonsequential
Discrete model				
Classical estimation	Complete	No	No	Both
Kalman filtering	Complete	No	No	Sequential

[a] After Nieman *et al.* [13].
[b] N/A denotes not appropriate.

for any negative constant u it has all negative real poles. If one attempts a frequency response on this system with an oscillator control rod, for example, $n(t)$ exhibits a slow divergence. Despite these difficulties, practicing engineers frequently "throw caution to the wind" and usually get valid results for the overall closed-loop system since it normally has negative feedback of neutron multiplication. Similar identification techniques using noise signals for nuclear reactors are discussed by Balcomb *et al.* [14], Gyftopoulos and Hooper [15], and Klystra and Uhrig [16]. Sequential identification of nuclear reactor systems is discussed by Mohler and Shen [11, pp. 306–313].

Several excellent surveys on system identification have been published in recent years by Nieman *et al.* [13], Aström and Eykhoff [17], Balakrishnan and Peterka [18], Cuenod and Sage [19], and Sage and Melsa [20]. Some of the more powerful sequential methods for nonlinear processes, which are subclasses of those listed in Table 1.1, are quasi-linearization, invariant embedding, gradient techniques, and ε-techniques. The direct differential method, consisting of the obvious algebra, is the most straightforward of the identification methods. Unfortunately, it requires a measurement or estimation of $\dot{\mathbf{x}}$ which is normally quite noisy. This method is used in Chapter V in conjunction with physiological modeling.

Consider a simple example to illustrate the application of the direct differential method with

$$d\mathbf{x}/dt = \mathbf{f}(\mathbf{x}, \mathbf{u}, \boldsymbol{\alpha}),$$

where $\boldsymbol{\alpha}$ is an unknown vector parameter; $\boldsymbol{\alpha}$ may be selected from measurements on $[t_0, t_f]$ so as to minimize

$$J(\boldsymbol{\alpha}) = \int_{t_0}^{t_f} \|\dot{\mathbf{x}} - \mathbf{f}(\mathbf{x}, \mathbf{u}, \boldsymbol{\alpha})\|^2 \, dt.$$

Here, $\|\cdot\|$ may be taken as the Euclidean norm. Hence it is necessary that $\partial J/\partial\boldsymbol{\alpha} = \mathbf{0}$, or

$$\int_{t_0}^{t_f} \frac{\partial \mathbf{f}}{\partial \boldsymbol{\alpha}} \cdot \dot{\mathbf{x}} \, dt = \int_{t_0}^{t_f} \frac{\partial \mathbf{f}}{\partial \boldsymbol{\alpha}} \cdot \mathbf{f} \, dt.$$

For the scalar bilinear system

$$dx/dt = ax + bux + cu,$$

with x, $\dot{x}$, u known on $[t_0, t_f]$ and a, c given constants,

$$b = \int_{t_0}^{t_f} u(x\dot{x} - ax^2\dot{x} - cux)\,dt \Big/ \int_{t_0}^{t_f} u^2x^2\,dt.$$

Sometimes it is convenient to define an extended state vector

$$\mathbf{z}(t) = \begin{bmatrix} \mathbf{x}(t) \\ \boldsymbol{\alpha} \end{bmatrix},$$

where the extended state equation is

$$d\mathbf{z}/dt = \begin{bmatrix} \dot{\mathbf{x}} \\ \mathbf{0} \end{bmatrix}.$$

In this manner, the problem of parameter identification becomes one of state identification. At this point it should be noted, however, that the extended system is not bilinear.

Bruni *et al.* [21] replace the extended state equation by appropriately transformed integral equations in order to analyze identification of bilinear systems. They show that if the measurements are assumed to be noise free, then the sequence of solutions generated from a quasi-linearization algorithm does converge to the actual parameter and state vector under quite general conditions which may readily be checked. While quasi-linearization may be applied directly to the differential bilinear model (1.9), the resulting conditions of convergence are not as apparent as for the integral equation model.

In order to consider random inputs, for which in some cases (e.g., white noise) the state is not differentiable for the differential bilinear model (1.9), Bruni *et al.* [21] utilize a stochastic integral equation. They also consider additive noise on the output measurements, which are linear in state, and show that the identification problem may be solved by a maximum likelihood algorithm applied to a sequence of Cauchy polygonal approximations to the solution of the transformed stochastic integral equation.

Balakrishnan [22] considers the parameter identification problem for bilinear systems with both additive state noise and additive output noise, but without input noise. For this case, linear system theory is appropriate, and Balakrishnan shows that a sequential maximum likelihood algorithm again converges under certain conditions.

1.4 VOLTERRA SERIES ANALYSIS

In general, system outputs may not necessarily be the state variables, and it may be more convenient to estimate system parameters directly from these outputs rather than indirectly from estimated states. Here it is assumed that the output is expressed by

$$\mathbf{y} = \mathbf{D}\mathbf{x}, \tag{1.11}$$

where $\mathbf{y} \in R^q$. Further, it is assumed that the system defined by (1.9) and (1.11) is at least locally controllable and locally observable in the neighborhood of $\mathbf{x} = \mathbf{0}$, $u = \mathbf{0}$ [23]. Then the identification process is concerned only with minimum system realizations.

The output of the nonlinear system may be related to the input by a functional series of kernels. For multivariable systems, this Volterra series has the form of

$$\mathbf{y}(t) = \sum_{i=1}^{\infty} \underbrace{\int_0^{\infty}\int_0^{\infty}\cdots\int_0^{\infty}}_{i} \sum_{k_1=1}^{m}\cdots\sum_{k_i=1}^{m} \mathbf{w}_i^{(k_1,\ldots,k_i)}(\tau_1,\ldots,\tau_i)$$
$$\cdot\mathbf{u}_{k_1}(t-\tau_1)\cdots\mathbf{u}_{k_i}(t-\tau_i)\,d\tau_1\cdots d\tau_i, \tag{1.12}$$

where $\mathbf{u}(t) = 0$ for $t < 0$. So that in general an infinite number of q-vector kernels $\mathbf{w}_i^{(k_1,\ldots,k_i)}(\tau_1,\ldots,\tau_i)$ are required. Frequently, the system may be approximated, however, by a finite-memory system. In other words, (1.12) may be approximated by a finite summation, and the integrations may be approximated by a finite summation.

The purpose of this section is to analyze the identification of bilinear system parameters in (1.9) and (1.11) from (1.12). Bruni *et al.* [24] show that the solution to (1.9) and (1.11) can be expressed as the limit of a uniformly converging sequence of solutions on $[0, t_f]$,

$$\lim_{i\to\infty} \mathbf{x}_i(t) = \mathbf{x}(t),$$

to linear systems of the form

$$\begin{aligned}\dot{\mathbf{x}}_0 &= \mathbf{A}\mathbf{x}_0 + \mathbf{C}\mathbf{u}\\ \dot{\mathbf{x}}_i &= \mathbf{A}\mathbf{x}_i + \sum_{k=1}^{m}\mathbf{B}_k u_k \mathbf{x}_{i-1} + \mathbf{C}\mathbf{u},\end{aligned} \tag{1.13}$$

where $i = 1, \ldots,$ and $\mathbf{x}_i(0) = \mathbf{x}_0$.

Proceeding according to Bruni *et al.* [24] analytical solutions are obtained for

$$\mathbf{x}_0(t) = e^{\mathbf{A}t}\mathbf{x}_0 + \int_0^t e^{\mathbf{A}(t-\sigma)}\mathbf{C}\mathbf{u}(\sigma)\,d\sigma$$

$$\mathbf{x}_i(t) - \mathbf{x}_{i-1}(t) = \int_0^t e^{\mathbf{A}(t-\sigma)} \sum_{k=1}^{m} \mathbf{B}_k u_k(\sigma)[\mathbf{x}_{i-1}(\sigma) - \mathbf{x}_{i-2}(\sigma)]\,d\sigma \tag{1.14}$$

$$i = 1, 2, \ldots; \mathbf{x}_{-1}(t) = \mathbf{0}.$$

Then by successive substitutions of (1.13) into

$$\mathbf{x}(t) = \sum_{i=0}^{\infty} \mathbf{x}_i(t) - \mathbf{x}_{i-1}(t),$$

with

$$\mathbf{C} = [\mathbf{c}_1 \cdots \mathbf{c}_m],$$

and upper limits of integration extended to the same t by multiplication of the integrands in (1.14) by an appropriate step function or generalized function, $\delta_{-1}(\sigma_i - \sigma_{i+1})$, it is seen that

$$\mathbf{x}(t) = \sum_{i=0}^{\infty} \underbrace{\int_0^t \cdots \int_0^t}_{i} \exp[\mathbf{A}(t - \sigma_1)] \sum_{k_1=1}^{m} \mathbf{B}_{k_1} \exp[\mathbf{A}(\sigma_1 - \sigma_2)] \sum_{k_2=1}^{m} \mathbf{B}_{k_2}$$
$$\cdots \exp[\mathbf{A}(\sigma_{i-1} - \sigma_i)] \sum_{k_i=1}^{m} \mathbf{B}_{k_i} \exp(\mathbf{A}\sigma_i)\mathbf{c}_{k_i}\delta_{-1}(\sigma_1 - \sigma_2)$$
$$\times\ \delta_{-1}(\sigma_2 - \sigma_3) \cdots \delta_{-1}(\sigma_{i-1} - \sigma_i) u_{k_1}(\sigma_1) u_{k_2}(\sigma_2)$$
$$\cdots u_{k_{i+1}}(\sigma_{i+1})\,d\sigma_1\,d\sigma_2 \cdots d\sigma_{i+1}. \tag{1.15}$$

Examination of the structure of (1.15), however, leads to further simplification with the state a sum of two terms, only one of which depends on the initial state $\mathbf{x}_0$ [24]. With zero initial state, $\mathbf{x}_0 = \mathbf{0}$, Eqs. (1.11) and (1.15) yield

$$\mathbf{y}(t) = \sum_{i=1}^{\infty} \underbrace{\int_0^t \int_0^t \cdots \int_0^t}_{i} \sum_{k_1=1}^{m} \cdots \sum_{k_i=1}^{m} \mathbf{D} \exp(\mathbf{A}\tau_1)\mathbf{B}_{k1} \exp[\mathbf{A}(\tau_2 - \tau_1)]\mathbf{B}_{k_2}$$
$$\cdots \mathbf{B}_{k_{i-1}} \exp[\mathbf{A}(\tau_i - \tau_{i-1})]\mathbf{c}_{k_i}\delta_{-1}(\tau_2 - \tau_1)\delta_{-1}(\tau_3 - \tau_2)$$
$$\cdots \delta_{-1}(\tau_i - \tau_{i-1}) u_{k_1}(t - \tau_1) u_{k_2}(t - \tau_2)$$
$$\cdots u_{k_i}(t - \tau_i)\,d\tau_1\,d\tau_2 \cdots d\tau_i, \tag{1.16}$$

where $\tau_k = t - \sigma_k$, $k = 1, 2, \ldots, i$. By comparison of (1.16) with (1.12) it is apparent that the Volterra kernels may be related to the bilinear system structure and the state transition matrix $e^{\mathbf{A}t}$ associated with the homogeneous part of (1.9). Consequently,

$$\begin{aligned}\mathbf{w}_i^{(k_1,\ldots,k_i)}(\tau_1,\ldots,\tau_i) = {} & \mathbf{D}\exp(\mathbf{A}\tau_1)\mathbf{B}_{k_1}\exp[\mathbf{A}(\tau_2-\tau_1)]\mathbf{B}_{k_2} \\ & \cdots \mathbf{B}_{k_{i-1}}\exp[\mathbf{A}(\tau_i-\tau_{i-1})]\mathbf{c}_{k_i}\delta_{-1}(\tau_1) \\ & \times \delta_{-1}(\tau_2-\tau_1)\cdots\delta_{-1}(\tau_i-\tau_{i-1}), \qquad (1.17)\\ i = 1, 2, \ldots; \quad k_1 = 1, 2, \ldots, m; & \quad \ldots; \quad k_i = 1, 2, \ldots, m.\end{aligned}$$

From (1.17) it is obvious that the ith-order kernel is nonzero only on the domain space R^i described by the set $S_i = \{\tau_1, \tau_2, \ldots, \tau_i\}$ such that

$$\tau_i \geq \tau_{i-1} \geq \cdots \geq \tau_2 \geq \tau_1 \geq 0.$$

The nonzero kernels evaluated on this domain are

$$\mathbf{w}_i^{(k_1,\ldots,k_i)}(\tau_1,\ldots,\tau_i) = \mathbf{D}\left(\prod_{j=1}^{i-1}\exp(\mathbf{A}\tau_j)\mathbf{B}_{k_j}\exp(-\mathbf{A}\tau_j)\right)\exp(\mathbf{A}\tau_i)\mathbf{c}_{k_i}. \qquad (1.18)$$

From (1.17) and (1.9) it is seen that the first-order kernels form the columns of the impulse response matrix for the linear system (1.2), which is obtained from (1.9) for a zero $\mathbf{B}_k$ matrix. In this manner, the linear system impulse response matrix is given by

$$\mathbf{W}_1(\tau_1) = [\mathbf{w}_1^{(1)}(\tau_1)\cdots\mathbf{w}_1^{(m)}(\tau_1)] = \mathbf{D}\exp(\mathbf{A}\tau_1)\mathbf{C}\delta_{-1}(\tau_1). \qquad (1.19)$$

Again, it is seen that linear system techniques of identification can be quite useful for bilinear systems if used in the proper manner. The $(q \times m)$ matrices of the second-order kernels obtained from (1.17) are given by

$$\begin{aligned}\mathbf{W}_2{}^k(\tau_1,\tau_2) &= [\mathbf{w}_2^{(k,1)}(\tau_1,\tau_2)\cdots\mathbf{w}_2^{(k,m)}(\tau_1,\tau_2)] \qquad (1.20)\\ &= \mathbf{D}\exp(\mathbf{A}\tau_1)\mathbf{B}_k\exp[\mathbf{A}(\tau_2-\tau_1)]\mathbf{C}\delta_{-1}(\tau_1)\delta_{-1}(\tau_2-\tau_1),\\ k &= 1, 2, \ldots, m.\end{aligned}$$

In order to identify the bilinear system matrices, $\mathbf{B}_k$ $(k = 1, \ldots, m)$, it is convenient to define

$$\mathbf{G} = \int_{t_1}^{t_2} [\mathbf{D}e^{\mathbf{A}\tau}]^{\mathrm{T}}\mathbf{D}e^{\mathbf{A}\tau}\, d\tau \tag{1.21}$$

$$\mathbf{H} = \int_{t_1}^{t_2} e^{\mathbf{A}\tau}\mathbf{C}[e^{\mathbf{A}\tau}\mathbf{C}]^{\mathrm{T}}\, d\tau, \tag{1.22}$$

and

$$\mathbf{M}_k = \int_{t_1}^{t_2}\int_{t_1+\tau}^{t_2+\tau} [\mathbf{D}\exp(\mathbf{A}\tau_1)]^{\mathrm{T}}\mathbf{w}_2{}^k(\tau_1, \tau_2)\{\exp[\mathbf{A}(\tau_2 - \tau_1)]\mathbf{C}\}^{\mathrm{T}}\, d\tau_2\, d\tau_1, \tag{1.23}$$

where T indicates a transpose operation. By replacing τ_2 by $\tau_1 + \theta$ in (1.23) for algebraic convenience, (1.20) yields

$$\mathbf{B}_k = \mathbf{G}^{-1}\mathbf{M}_k\mathbf{H}^{-1}, \qquad k = 1, \ldots, m. \tag{1.24}$$

It is convenient and enlightening to study the bilinear system (1.9) with scalar control, $m = 1$, and scalar output, $q = 1$. Then the ensuing Volterra series and system realization follow directly from (1.16) to (1.24) with $\mathbf{c} \in R^n$ replacing $\mathbf{C}$ and $\mathbf{d}^{\mathrm{T}} \in R^n$ replacing $\mathbf{D}$.

The kernels $w_i(\tau_1, \ldots, \tau_i)$ in the Volterra series can be replaced by symmetrical ones, $v_i(\tau_1, \ldots, \tau_i)$, such that

$$v_i(\tau_1, \ldots, \tau_i) = \sum_{\text{per}} w_i(\tau_1, \ldots, \tau_i), \tag{1.25}$$

where the summation is carried out over all $i!$ permutations of $\tau_1, \ldots, \tau_i$.

Algebraically, it is further convenient to decompose the output $y(t)$ into the sum

$$y(t) = y_x(t) + y_u(t) + y_{xu}(t), \tag{1.26}$$

where $y_x(t) = \mathbf{d}^{\mathrm{T}}e^{\mathbf{A}t}\mathbf{x}_0$ is the zero-input response, and $y_u(t)$ is the zero-state response [25]. There, the zero-state response is equal to

$$y_u(t) = \sum_{i=1}^{\infty} (1/i!) \int_0^t v_i(\tau_1, \ldots, \tau_i)\left[\prod_{k=1}^{i} u(t - \tau_k)\right] d\tau_1 \cdots d\tau_k \tag{1.27}$$

with $v_i(\tau_1, \ldots, \tau_i)$ computed from (1.25) and (1.18). The remaining response $y_{xu}(t)$ with nonzero initial state can be expanded in a similar

Volterra series:

$$y_{xu}(t) = \sum_{i=1}^{\infty} (1/i!) \int_0^t z_i(\tau_1, \ldots, \tau_i)\mathbf{B} \exp[\mathbf{A}(t - \tau_i)]\mathbf{x}_0 \times \left[\prod_{k=1}^{i} u(t - \tau_k)\right] d\tau_1 \cdots d\tau_i, \quad (1.28)$$

where $z_i(\tau_1, \ldots, \tau_i)$ may be computed from (1.25), (1.18), and (1.28) with $z_i(\tau_1, \ldots, \tau_i)$ replacing $v_i(\tau_1, \ldots, \tau_i)$ and with $\mathbf{c}$ replaced by the appropriate identity matrix [25].

The sequence of symmetric kernels $\{v_i(\tau_1, \ldots, \tau_i)\}_1^\infty$ is said to be realizable by means of a bilinear system (1.19) if and only if there exist four constant matrices $\mathbf{A}$, $\mathbf{B}$, $\mathbf{c}$, $\mathbf{d}$ such that (1.18) and (1.25) are satisfied. From (1.18) and (1.25) it is readily apparent that (i) $w_1(\tau_1)$ is a linear system impulse response with proper rational Laplace transform, and that (ii) $v_i(\tau_1, \ldots, \tau_i)$ for all $i > 1$, may be factored into similar functions with proper rational Laplace transforms. While (1.18) and the ensuing results were obtained only for the time domain defined by S_i, the use of symmetric kernels implies that the bilinear realizability condition obtained from (1.18) and (1.25) is satisfied for all time [25].

In terms of these factors, $v_i(\tau_1, \ldots, \tau_i)$ has the form of

$$v_i(\tau_1, \ldots, \tau_i) = \mathbf{H}(\tau_1)\mathbf{F}(\tau_2 - \tau_1) \cdots \mathbf{F}(\tau_{i-1} - \tau_{i-2})\mathbf{G}(\tau_i - \tau_{i-1}), \quad (1.29)$$

where $\mathbf{F}(t)$ is $s \times s$, $\mathbf{G}(t)$ is $s \times 1$, and $\mathbf{H}(t)$ is $1 \times s$. If $\mathbf{B}$ is factored in the form

$$\mathbf{B} = \mathbf{B}^1\mathbf{B}^2,$$

where $\mathbf{B}^1$ is $n \times s$ and $\mathbf{B}^2$ is $s \times n$, then the three characteristic matrices are described by

$$\mathbf{H}(t) = \mathbf{d}^{\mathrm{T}} e^{\mathbf{A}t}\mathbf{B}^1$$

$$\mathbf{F}(t) = \mathbf{B}^2 e^{\mathbf{A}t}\mathbf{B}^1$$

$$\mathbf{G}(t) = \mathbf{B}^2 e^{\mathbf{A}t}\mathbf{c}.$$

It is apparent that all bilinear realizations can be otained from all the linear realizations that are associated with all the matrices

$$\mathbf{L}(t) = \begin{bmatrix} \mathbf{w}_1(t) & \mathbf{H}(t) \\ \mathbf{G}(t) & \mathbf{F}(t) \end{bmatrix}$$

relating to the given sequence of kernels [25]. From this, d'Alessandro *et al.* [25] show that the minimal bilinear realization is such that:

(a) the state space dimension is given by

$$n_0 = \text{rank}\begin{bmatrix} \mathbf{v}_1(t) & \mathbf{H}_0(t) \\ \mathbf{G}_0(t) & \mathbf{F}_0(t) \end{bmatrix},$$

where $\{\mathbf{F}_0(t), \mathbf{G}_0(t), \mathbf{H}_0(t)\}$ is a minimal factorization (i.e., most reduced form) of the subsequence $\{v_i(t_1, \ldots, t_i)\}_2^\infty$; and

(b) the matrix $\mathbf{B}$ has minimum rank given by

$$r_0 = m_0 \equiv \dim\{\mathbf{F}_0(t), \mathbf{G}_0(t), \mathbf{H}_0(t)\},$$

the dimension of the minimal factorization. Consequently, the minimal realization depends on the minimum number of necessary bilinear multipliers.

D'Alessandro *et al.* [25] also show that a factorized sequence of kernels $\{v_i(t_1, \ldots, t_i)\}_2^\infty$ is uniquely determined by the finite sequence $\{v_i(t_1, \ldots, t_i)\}_2^{2m_0+1}$. This is a most important property of bilinear realizations.

From this work, it is seen that minimal linear system realization procedures may be applied to minimal bilinear realizations by proper structural decompositions. It is significant that a finite number of kernels is sufficient to realize the finite-dimensional bilinear system from its Volterra series.

The identification process is summarized here. The first step provides a realization of $\mathbf{A}$, $\mathbf{C}$, $\mathbf{D}$ from classical procedures for the linear system's impulse response matrix, which is composed of the m first-order q-vector kernels. Linear system realization, which may be applied here, is discussed by Kalman [26, 27]. The second step is given by (1.24). As usual, there are possible trouble spots. It is readily seen that both steps deduce realizations only up to a linear constant transformation of the state space. The order of the minimal realization obtained in the first step might not be equal to the order n of the bilinear system, which depends also on the higher order kernels. Also, it must be realized that measurements made in practice involve noise, and statistical estimation of the required terms then becomes essential.

Exercises

1.1 Test the Delphi method for a common problem with multiple locally optimal solutions on a group of friends.

1.2 (a) Derive the neutron fission transfer function (1.10) from (1.4) and (1.5) for perturbations about n_0 and $u_0 = 1$.
(b) Explain why such a system is said to be structurally unstable.

1.3 Explain the variable structure of the bilinear system (1.9) and Fig. 1.3 as compared to the rigid structure of the linear system.

1.4 Show that the first-order kernel of the Volterra series characterizes the linear portion of the bilinear system.

1.5 Derive the Volterra kernels to describe the dynamical behavior of the braking car (1.8). Explain the number of symmetric kernels required.

1.6 Why are controllability and observability important for system identification? Compare their complications for bilinear systems as compared to linear systems.

1.7 Show that linear system techniques may be applied to the realization of bilinear systems.

1.8 (a) How does control measurement noise considerably complicate parameter identification for bilinear systems compared to that of liner systems?
(b) Can the same be said about additive state noise? Why?

REFERENCES

1. Weed, L. L., "Medical Records, Medical Education and Patient Care." Case-Western Res. Univ. Press, Cleveland, Ohio, 1964.
2. Davis, R. M., Communication for the medical community—a prototype of a special interest audience. *Proc. AIAA Annu. Meeting, 6th, Anaheim, California, October 20–24, 1969*, Paper No. 69-1072.
3. Interdisciplinary research topics in urban engineering. Rep. Urban Engrg. Study Committee. Amer. Soc. Engrg. Education, Washington, D.C., 1969.
4. Future fields of control application. *NASA Symp. MIT, Cambridge, Massachusetts, February 10–11, 1969*. Available through Sci. and Tech. Information Div., Office of Technol. Utilization, NASA, Washington, D.C., 1969.

5. Hsu, J. C., and Meyer, A. V., "Modern Control Principles and Applications." McGraw-Hill, New York, 1968.
6. Coddington, E. A., and Levinson, N., "Theory of Ordinary Differential Equations." McGraw-Hill, New York, 1955.
7. Bellman, R. E., "Stability Theory of Differential Equations." McGraw-Hill, New York, 1953.
8. LaSalle, J., and Lefschetz, S., "Stability by Liapunov's Direct Method with Applications." Academic Press, New York, 1961.
9. Aizerman, M. A., and Gantmacher, F. R., "Absolute Stability of Regulatory Systems." Holden-Day, San Francisco, California, 1964.
10. Zadeh, L. A., and Desoer, C. A., "Linear System Theory." McGraw-Hill, New York, 1963.
11. Mohler, R. R., and Shen, C. N., "Optimal Control of Nuclear Reactors." Academic Press, New York, 1970.
12. Keepin, G. R., "Physics of Nuclear Kinetics." Addison-Wesley, Reading, Massachusetts, 1965.
13. Nieman, R. E., Fisher, D. G., and Seborg, D. E., A review of process identification and parameter estimation techniques. *Internat. J. Control* **13**, 209–264 (1971).
14. Balcomb, J. D., Demuth, H. B., and Gyftopoulos, E. P., A cross-correlation method for measuring the impulse response of reactor systems. *Nuclear Sci. Engrg.* **11**, 159–166 (1961).
15. Gyftopoulos, E. P., and Hooper, R. J., Signals for transfer-function measurements in non-linear systems. Noise analysis in nuclear systems. *AEC Symp. Ser.* **4**, 335–345 (1964).
16. Klystra, C. D., and Uhrig, R. E., Measurement of the spatially dependent transfer function. Noise analysis in nuclear systems. *AEC Symp. Ser.* **4**, 285–320 (1964).
17. Aström, K. J., and Eykhoff, P., System identification—A survey. *Automatica J. IFAC* **7**, 123–163 (1971).
18. Balakrishnan, A. V., and Peterka, V., Identification in automatic control systems. *Automatica J. IFAC* **5**, 817–829 (1969).
19. Cuenod, M., and Sage, A., Comparison of some methods used for process identification. *Automatica J. IFAC* **4**, 235–269 (1969).
20. Sage, A., and Melsa, J. L., "System Identification." Academic Press, New York, 1971.
21. Bruni, C., DiPillo, G., and Koch, G., Mathematical models and identification of bilinear systems. *In* "Theory and Application of Variable Structure Systems" (R. R. Mohler and A. Ruberti, eds.). Academic Press, New York, 1972.
22. Balakrishnan, A. V., Modeling and identification theory. *In* "Theory and Application of Variable Structure System" (R. R. Mohler and A. Ruberti, eds.). Academic Press, New York, 1972.
23. Lee, E. B., and Markus, L., "Foundations of Optimal Control Theory." Wiley, New York, 1967.

24. Bruni, C., DiPillo, G., and Koch, G., On the mathematical models of bilinear systems. *Ricerche Automat.* **2**, No. 1 (1971).
25. d'Alessandro, P., Isidori, A., and Ruberti, A., Realization and structure theory of bilinear dynamical systems. *In* "Theory and Application of Variable Structure Systems" (R. R. Mohler and A. Ruberti, eds.). Academic Press, New York, 1972.
26. Kalman, R. E., Irreducible realization and the degree of a matrix of rational functions. *SIAM J. Appl. Math.* **13**, 520–544 (1965).
27. Kalman, R. E., Mathematical description of linear dynamical systems. *SIAM J. Control* **1**, 152–192 (1963).

II

Controllability

2.1 BACKGROUND

2.1.1 Concept and Definition

Controllability is a basic property of systems which is indicative of the ability to control. In Chapter I a form of controllability is assumed in regard to minimal realizations of system order. Similarly, in Chapter III, this fundamental concept is necessary to form meaningful optimal control problems. Controllability of nuclear reactors and the reachable zone of a physiological bilinear system are studied in Chapters IV and V, respectively. While controllability and optimal control usually are recognized as characteristics of modern control theory, Caratheodory [1] established some base in these areas several decades ago. In his theory on thermodynamics, he presented the concept of controllability and studied the reachable zones of nonlinear thermodynamic processes with multiplicative control.

There are numerous degrees of state controllability and output controllability that are formally defined in the literature. These are discussed in detail by Hsu and Meyer [2], Athans and Falb [3], and Lee and Markus [4]. In this book, controllability will refer to state controllability unless otherwise specified. This is the most common convention.

A most harsh requirement of a system would be to specify that it must be completely controllable. Formally, a system is said to be *completely controllable* if it can be driven from any finite initial state $\mathbf{x}_0 \in R^n$ to any prescribed finite terminal state $\mathbf{x}_f \in R^n$ in finite time with an admissible control.

2.1.2 Linear Systems and Their Controllability Limitations

Numerous results [2–4] are published on controllability of linear systems which stem mainly from the work of Kalman. (See Kalman *et al.* [5].) Probably the most important single theorem from this work is the following:

Theorem 1 The linear time-invariant system [as defined by Eq. (1.4) and repeated for convenience]

$$d\mathbf{x}/dt = \mathbf{Ax} + \mathbf{Cu}, \tag{2.1}$$

where $\mathbf{x} \in R^n$ and $\mathbf{u} \in R^m$ is completely controllable if and only if the $n \times nm$ matrix

$$\mathbf{E} = [\mathbf{C} \mid \mathbf{AC} \mid \cdots \mid \mathbf{A}^{n-1}\mathbf{C}] \tag{2.2}$$

has rank n.

It is significant to note, however, that complete controllability of linear systems usually requires infinite control—a condition that is not physically realizable in practice. In fact, it is easy to prove the following theorem:

Theorem 2 The linear system (2.1) with $\mathbf{u}$ bounded is *not* completely controllable if the eigenvalues of $\mathbf{A}$ have negative real parts.

To show this, first consider the autonomous system

$$d\mathbf{x}/dt = \mathbf{Ax}, \tag{2.3}$$

where $\mathbf{A}$ has negative real eigenvalues. Since (2.3) is asymptotically stable, there exists a positive-definite quadratic form (a Liapunov function) $V(\mathbf{x})$ with $\dot{V}(\mathbf{x})$ a negative-definite quadratic form $q(\mathbf{x})$. Then the surfaces $V(\mathbf{x}) =$ constant are hyperspheroids which enclose the origin in state space.

The same system with an added bounded linear control term is defined by (1.4). By taking the inner product of the gradient of V with the state velocity it is seen that

$$\dot{\mathbf{x}}\cdot\frac{\partial V}{\partial \mathbf{x}} = \mathbf{Ax}\cdot\frac{\partial V}{\partial \mathbf{x}} + \mathbf{Cu}\cdot\frac{\partial V}{\partial \mathbf{x}}$$

$$= q(\mathbf{x}) + \mathbf{Cu}\cdot\frac{\partial V}{\partial \mathbf{x}}, \qquad (2.4)$$

where $q(\mathbf{x})$ is negative definite and the last term is linear in $\mathbf{x}$. Now, (2.4) is negative for every bounded $\mathbf{u}$ with $\mathbf{x}$, and therefore $V(\mathbf{x}) = M$ sufficiently large. Hence the state velocity vector $\dot{\mathbf{x}}$ points into the interior of the volume which is bounded by $V(\mathbf{x}) = M$, and the stable linear system is not completely controllable with bounded control.

This short proof suggests that there is some fundamental relationship between stability and controllability. In Section 2.3, this relationship is further exploited to provide sufficient conditions for complete controllability of bilinear systems.

It is of interest to note that the linear harmonic oscillator does not fit the above-mentioned class of linear systems lacking controllability. It is readily seen, for example, that the second-order linear system with imaginary eigenvalues is completely controllable with bounded control if a sufficiently large number of switchings are permitted with a bang–bang control [4].

2.2 CONTROLLABLE BILINEAR SYSTEMS

The variable structure of bilinear systems allows them to be more controllable than linear systems just as it frequently provides for a more accurate model. In control system design, it may be convenient or even necessary to introduce adaptive or variable-structure control to steer the plant throughout the required region of state space. Complete controllability already has been introduced as a useful concept from a mathematical standpoint. While complete controllability can never be realized in practice due to physical constraints, there are numerous pursuit systems, such as the TFX aircraft or the ABM, for which it may be necessary to attain relatively large regions of state space.

2.2.1 Bilinear and Linear Controllability Comparisons

The following example shows how a linear system that is not completely controllable with bounded control may be made so by the addition of a bilinear mode of control.

Example 1 *Bilinear control synthesis for complete controllability.* Consider the system

$$\begin{aligned} dx_1/dt &= x_2 \\ dx_2/dt &= -2x_1 - x_2 + u. \end{aligned} \tag{2.5}$$

First, with unbounded control $u(t)$, it is seen that the rank of $\mathbf{E}$ given by (2.2) is equal to 2, the system order; hence the system is completely controllable if infinite control is allowed. If the control is constrained, however, so that $|u| < 1$, it is readily seen from the theorem above that the system is not completely controllable. It is apparent from the so-called bang–bang principle [4] and the characteristic stable-node state behavior for the homogeneous part of (2.5) that the constrained system does have a degree of controllability or null controllability [4] such that any state between the lines $x_1 = -1$ and $x_1 = +1$ on the (x_1, x_2) state plane may be driven to the origin in finite time.

If an appropriate bilinear control is added to (2.5), however, the system can be made completely controllable. For example, consider

$$\begin{aligned} \dot{x}_1 &= x_2 \\ \dot{x}_2 &= -2x_1 - x_2 + u + x_1u + 2x_2u, \end{aligned} \tag{2.6}$$

with $|u| \leq 1$. For $u = +1$, the system exhibits an unstable-focus characteristic with an equilibrium point at $(1, 0)$ on the phase plane, and for $u = -1$ a stable focus with equilibrium point at $(-\frac{1}{3}, 0)$. The corresponding eigenvalues are $\lambda_1, \lambda_2 = \frac{1}{2}(1 \pm j\sqrt{3})$ and $\lambda_1, \lambda_2 = -(\frac{1}{2})(3 \pm j\sqrt{3})$, respectively. It is apparent that such combinations of stable and unstable foci are completely state controllable. In general, if the bilinear control which is added to the second equation of (2.5) has the form

$$b_{21}x_1u + b_{22}x_2u,$$

then an evaluation of the system eigenvalues with $u = +1$ and $u = -1$ shows that a combination of stable and unstable foci, thus ensuring complete controllability, exists if $b_{22} > 1$ and $b_{21} < \frac{1}{2}(b_{22} + 4)$.

This example shows that roughly speaking, additive linear control may be used to control to new equilibrium states while multiplicative linear control may be used to vary the structure and therefore the range of attainable states—a property that is exploited more precisely in Section 2.3.

The following example shows that, as should be expected, not every bilinear system is completely controllable.

Example 2 *Saddle-point superposition.* The linear system

$$\begin{aligned} dx_1/dt &= x_2 \\ dx_2/dt &= x_1 - x_2 + u \end{aligned} \tag{2.7}$$

is obviously completely controllable with infinite control allowed since the rank of $\mathbf{E}$ from (2.2) is $n = 2$. Even though the linear system is unstable and does not fit the previous theorem, the system is not completely controllable with bounded control. In this case there is a region of null controllability between two parallel lines corresponding to the stable eigenvectors for the saddle point at $(-1, 0)$ and at $(+1, 0)$ with $u = +1$ and $u = -1$, respectively. If the bilinear control $-x_2 u$ is added to the second equation of (2.7), the system still exhibits saddle-point portraits for $u = +1$ and $u = -1$. Again, the superposition of these state-plane portraits will yield the region of controllability—in this case not the entire state space. Again, however, it is possible to show complete controllability by a superposition of stable- and unstable-foci portraits for $u = +1$ and $u = -1$ with an appropriate bilinear control.

2.2.2 Independent Linear and Bilinear Controls

Several conclusions are readily apparent about controllability of the special bilinear system

$$d\mathbf{x}/dt = \mathbf{A}\mathbf{x} + u\mathbf{B}\mathbf{x} + \mathbf{c}\mathbf{v}, \tag{2.8}$$

where u is a scalar control, and $\mathbf{v} \in R^{m-1}$ and is independent of u. The other parameters are defined as before. Here it is assumed that $|u| \leq 1$, $\mathbf{v}$ is unconstrained, and both are at least piecewise continuous again. It is seen from the work of Kalman *et al.* [6] that (2.8) is not completely controllable if there exists a nonsingular linear transformation such that

$\mathbf{x} = \mathbf{T}\mathbf{y}$ transforms (2.8) into the form

$$\begin{bmatrix} d\mathbf{y}_1/dt \\ d\mathbf{y}_2/dt \end{bmatrix} = \begin{bmatrix} \mathbf{A}^{11} & \mathbf{A}^{12} \\ \mathbf{A}^{21} & \mathbf{A}^{22} \end{bmatrix} \begin{bmatrix} \mathbf{y}_1 \\ \mathbf{y}_2 \end{bmatrix} + u \begin{bmatrix} \mathbf{B}^{11} & \mathbf{B}^{12} \\ \mathbf{B}^{21} & \mathbf{B}^{22} \end{bmatrix} \begin{bmatrix} \mathbf{y}_1 \\ \mathbf{y}_2 \end{bmatrix} + \begin{bmatrix} \mathbf{c}^1 \\ \mathbf{c}^2 \end{bmatrix} \mathbf{v}, \tag{2.9}$$

where $\mathbf{y}_1$, $\mathbf{y}_2$ are vectors of n_c controllable and $(n - n_c)$ noncontrollable components, respectively, and $\mathbf{A}^{21}, \mathbf{B}^{21}, \mathbf{c}^2$ are zero matrices of dimension $(n - n_c) \times n_c$, $(n - n_c) \times n_c$, and $(n - n_c) \times (m - 1)$, respectively, with $n - n_c > 0$. Obviously state trajectories for such systems cannot leave the linear subspace of dimension n_c once it has entered it. For example, the system

$$d\mathbf{x}/dt = \begin{bmatrix} 1 \\ 1 \end{bmatrix} v + u \begin{bmatrix} 1 & -1 \\ 0 & 0 \end{bmatrix} \mathbf{x} \tag{2.10}$$

is algebraically equivalent to

$$d\mathbf{y}/dt = \begin{bmatrix} 1 \\ 0 \end{bmatrix} v + u \begin{bmatrix} 0 & 0 \\ 0 & 1 \end{bmatrix} \mathbf{y} \tag{2.11}$$

by the transformation $\mathbf{x} = \mathbf{T}\mathbf{y}$, with

$$\mathbf{T} = \begin{bmatrix} 1 & 1 \\ 1 & 0 \end{bmatrix}.$$

Here, the $y_2 = 0$ axis is the subspace which can be reached from any exterior state, but no admissible control can cause a trajectory to leave it.

For bilinear systems of the form described by (2.8), there are several conclusions that can be derived directly from linear system theory. For example, (2.8) is not completely controllable if and only if $(\mathbf{A}, \mathbf{C})$ and $(\mathbf{B}, \mathbf{C})$ are not completely controllable. [Here, $(\mathbf{A}, \mathbf{C})$ refers to (2.8) with $\mathbf{B} = 0$ and similarly $(\mathbf{B}, \mathbf{C})$ refers to the system with $\mathbf{A} = 0$.] Then it is sufficient for either $(\mathbf{A}, \mathbf{C})$ or $(\mathbf{B}, \mathbf{C})$ to be completely controllable for the bilinear system (2.8) to be completely controllable. On the other hand, the general bilinear system (1.9) can be completely controllable even if $(\mathbf{A}, \mathbf{C})$ and $(\mathbf{B}, \mathbf{C})$ are not completely controllable.

Kučera [7] analyzes certain aspects of controllability of a bilinear system

$$d\mathbf{x}/dt = \sum_{k=1}^{\alpha} \mathbf{u}_k \mathbf{B}_k \mathbf{x} + \mathbf{C}\mathbf{v}, \tag{2.12}$$

where $\mathbf{v} \in R^{m-\alpha}$, $\mathbf{u} \in R^{\alpha}$, and the other terms are as defined in Eq. (1.9). Also, $\mathbf{u}$, $\mathbf{v}$ are bounded and measurable controls on $[0, \infty)$. The main result for this system, based on the theory of Lie groups as given by Chevalley [8], is to define the set of attainable points in state space by a so-called maximal integral manifold. Further, Kučera shows that the attainable states can be reached by piecewise constant controls with ordinates of only $-1, 0, 1$ for control constraints such that

$$\sum_{k=1}^{\alpha} |u_k| + \sum_{j=1}^{m-\alpha} |v_j| = 1.$$

Similar results are reported by Kučera [9] for the special bilinear system

$$d\mathbf{x}/dt = [\mathbf{A}(1 - u) + \mathbf{B}u]\mathbf{x}. \tag{2.13}$$

2.3 SUFFICIENT CONDITIONS FOR COMPLETE CONTROLLABILITY

Numerous necessary and sufficient conditions that are simple to apply have been derived for linear systems. The most popular of these utilizes the **E** matrix as discussed in Section 2.1. While linear systems are not generally completely controllable when the control is bounded, it was shown above that nonlinear systems can be designed to enhance controllability. Relatively few results, however, have been derived for nonlinear systems, and those tests that are available are generally not so convenient to apply. A good summary of work published prior to 1968 on controllability of nonlinear systems is given by Lee and Markus [4, pp. 264–394]. Kalman *et al.* [10] relate control theory to automata theory. In particular they extend controllability results to a class of additive systems which include linear systems. While the theory is powerful, the results are not directly applicable to bilinear systems.

In his early work, Caratheodory introduced Pfaffian equations which Hermes [11] shows are quite naturally associated with systems which are linear in control (for which the bilinear system is a special case). Hermes [11] then derives system controllability from integrability of the associated Pfaffian equations.

Sufficient conditions for complete controllability of bilinear systems are derived by Rink and Mohler [12], and the analysis presented next follows that work.

Again the bilinear system is given by

$$d\mathbf{x}/dt = \left(\mathbf{A} + \sum_{k=1}^{m} u_k \mathbf{B}_k\right)\mathbf{x} + \mathbf{Cu}, \tag{2.14}$$

where the parameters are as defined before for (1.9). Now, it is assumed that the class of admissible controls $\{\mathbf{u}(t)\}$ is the class of all piecewise continuous vector time functions with domain $[0, \infty)$ and range U, where U is a compact connected set containing the origin in R^m.

The reachable zone from an initial state $\mathbf{x}_0$, $R(\mathbf{x}_0) \subset R^n$, is the set of all states to which the system can be transferred in finite time, starting at $\mathbf{x}_0$. Similarly, the incident zone to a terminal state $\mathbf{x}_f$, $I(\mathbf{x}_f) \subset R^n$ is the set of all initial states from which $\mathbf{x}_f$ is reachable in finite time.

For each fixed $\mathbf{u} \in U$, the bilinear system is a constant-parameter linear system with system matrix $\mathbf{A} + \sum_{k=1}^{m} u_k \mathbf{B}_k$. The terms $\sum_{k=1}^{m} u_k \mathbf{B}_k$ in the system matrix permit manipulation of the eigenvalues of the fixed-control system. With an appropriate controller it is often possible to shift these eigenvalues from the left half of the complex plane to the right half-plane.

The controllability analysis presented here can be summarized by the following sufficient conditions.

Main Result The bilinear system (2.14) is completely controllable if:

(a) there exist control values $\mathbf{u}^+$ and $\mathbf{u}^-$ such that the real parts of the eigenvalues of the system matrix are positive and negative, respectively, and such that equilibrium states $\mathbf{x}_e(\mathbf{u}^+)$, $\mathbf{x}_e(\mathbf{u}^-)$ are contained in a connected component of the equilibrium set;

(b) for each $\mathbf{x}$ in the equilibrium set with an equilibrium control $\mathbf{u}_e(\mathbf{x}) \in U$ such that $\mathbf{f}(\mathbf{x}, \mathbf{u}_e(\mathbf{x})) = \mathbf{0}$, there exists a $\mathbf{v} \in R^m$ such that $\mathbf{g}$ lies in no invariant subspace of dimension at most $(n - 1)$ of the matrix $\mathbf{E}$, where

$$\mathbf{E} = \mathbf{A} + \sum_{k=1}^{m} u_k(\mathbf{x})\mathbf{B}_k \tag{2.15}$$

and

$$\mathbf{g} = \mathbf{Cv} - \sum_{l=1}^{m} \mathbf{v}_l\left[\mathbf{B}_l\left(\mathbf{A} + \sum_{k=1}^{m} u_k \mathbf{B}_k\right)^{-1} \mathbf{Cu}\right]. \tag{2.16}$$

Remark 1. For phase-variable systems, $x_1 = x, x_2 = \dot{x}, \ldots, x_n = x^{(n-1)}$, condition (b) is always satisfied if $\mathbf{C}$ is a nonzero matrix.

Remark 2. Condition (a) is satisfied if all the eigenvalues of the system matrix $\mathbf{A} + \sum_{k=1}^{m} u_k \mathbf{B}_k$ can be shifted across the imaginary axis of the complex plane without passing through zero, as $\mathbf{u}$ ranges continuously over a subset of U.

It is the objective of this section to substantiate these statements. First suppose there exists a fixed control value $\mathbf{u}^-$ in the interior of U such that the eigenvalues of the system matrix $\mathbf{A} + \sum_{k=1}^{m} u_k^- \mathbf{B}_k$ all have negative real parts. Then the trajectories of the system with constant control $\mathbf{u} = \mathbf{u}^-$ cover all of R^n, and each trajectory approaches the unique equilibrium state

$$\mathbf{x}_e(\mathbf{u}^-) = -\left(\mathbf{A} + \sum_{k=1}^{m} u_k^- \mathbf{B}_k\right)^{-1} \mathbf{C}\mathbf{u}^-. \tag{2.17}$$

For any initial state $\mathbf{x}_0 \in R^n$, the unique trajectory passing through $\mathbf{x}_0$ with control $\mathbf{u}^-$ reaches any neighborhood of $\mathbf{x}_e(\mathbf{u}^-)$ in finite time.

Suppose, also, that there exists a fixed control value $\mathbf{u}^+$ in the interior of U such that the eigenvalues of the matrix $\mathbf{A} + \sum_{k=1}^{m} u_k^+ \mathbf{B}_k$ all have positive real parts. Then the trajectories of the system with constant control $\mathbf{u} = \mathbf{u}^+$ cover all of the R^n, and each trajectory corresponds to motion away from the unique equilibrium state

$$\mathbf{x}_e(\mathbf{u}^+) = -\left(\mathbf{A} + \sum_{k=1}^{m} u_k^+ \mathbf{B}_k\right)^{-1} \mathbf{C}\mathbf{u}^+. \tag{2.18}$$

For any terminal state $\mathbf{x}_f$, the unique trajectory passing through $\mathbf{x}_f$ with control $\mathbf{u}^+$ reaches $\mathbf{x}_f$ from any neighborhood of $\mathbf{x}_e(\mathbf{u}^+)$ in finite time.

If $\mathbf{u}^-$ and $\mathbf{u}^+$ both exist, and if every point of some neighborhood of $\mathbf{x}_e(\mathbf{u}^+)$ can be reached from every point of some neighborhood of $\mathbf{x}_e(\mathbf{u}^-)$, then certainly any terminal state $\mathbf{x}_f$ can be reached from any initial state $\mathbf{x}_0$, and the system is certainly controllable. Such equilibrium sets and their connectedness are described in Section 2.4 for bilinear systems. Then the controllability analysis can be completed by means of a local controllability theorem due to Lee and Markus [13], which provides sufficient conditions for local controllability of a system in a neighborhood of its equilibrium set.

Theorem 3 Consider

$$\dot{\mathbf{x}} = \mathbf{f}(\mathbf{x}, \mathbf{u}),$$

where $\mathbf{f}(\mathbf{x}, \mathbf{u}) \in C^1$ in $R^n \times U$. If, for $\mathbf{x} \in R^n$, there exists an equilibrium control value $u_e(\mathbf{x})$ in the interior of U such that (i) $\mathbf{f}(\mathbf{x}, \mathbf{u}_e(\mathbf{x})) = \mathbf{0}$, and (ii) there exists a $\mathbf{v} \in R^m$ such that $\mathbf{Dv}$ lies in no invariant subspace of $\mathbf{E}$ of dimension at most $(n - 1)$, where

$$\mathbf{C} = \frac{\partial \mathbf{f}}{\partial \mathbf{u}}(\mathbf{x}, \mathbf{u}_e(\mathbf{x})) \qquad \text{and} \qquad \mathbf{E} = \frac{\partial \mathbf{f}}{\partial \mathbf{x}}(\mathbf{x}, \mathbf{u}_e(\mathbf{x}))$$

are real matrices, then $R(\mathbf{x})$ and $I(\mathbf{x})$ are open connected subsets of R^n.

If the equilibrium set is a connected proper subset of R^n and condition (ii) above is satisfied for every interior $\mathbf{u} \in U$, then the reachable zone from any equilibrium state $\mathbf{x}$ corresponding to an interior control value $\mathbf{u}_e(\mathbf{x})$ includes every other equilibrium state corresponding to an interior control value of U. For, if $R(\mathbf{x})$ has an equilibrium state $\mathbf{y}$ on its boundary, then $I(\mathbf{y})$ must be disjoint from $R(\mathbf{x})$. But, by the Lee–Markus theorem, $I(\mathbf{y})$ is an open neighborhood of $\mathbf{Y}$ if $\mathbf{u}_e(\mathbf{y})$ is in the interior of U. Hence $\mathbf{u}_e(\mathbf{y})$ must be on the boundary of U.

For the general bilinear system, the matrix $\mathbf{D}$ is

$$\mathbf{D} = \frac{\partial \mathbf{f}}{\partial \mathbf{u}}(\mathbf{x}, \mathbf{u}_e(\mathbf{x})) = [\mathbf{B}_1\mathbf{x} \mid \mathbf{B}_2\mathbf{x} \mid \cdots \mid \mathbf{B}_m\mathbf{x}] + \mathbf{C}, \tag{2.19}$$

and $\mathbf{E}$ is defined by (2.15). Substitution of the expression for the equilibrium state (see the derivation in the next section) into (2.15) yields the Lee–Markus theorem for a bilinear system: The system (2.14) is locally controllable at the equilibrium state corresponding to an interior $\mathbf{u} \in U$ if there exists a $\mathbf{v} \in R^m$ such that

$$-\sum_{j=1}^{m} \mathbf{v}_j \left[\mathbf{B}_j \left(\mathbf{A} + \sum_{k=1}^{m} u_k \mathbf{B}_k \right)^{-1} \mathbf{Cu} \right] + \mathbf{Cv}$$

lies in no invariant subspace of dimension at most $(n - 1)$ of the matrix $\mathbf{E}$.

The practical implications of using this criterion for a general bilinear system appear quite formidable. However, it will now be shown that the application to systems of the phase-variable type is straightforward, and in fact such systems always satisfy the criterion when $\mathbf{C} \neq \mathbf{0}$.

The matrix **D** for the phase-variable system is obtained by substituting the equilibrium-state expression (see Section 2.4) into (2.19). Hence,

$$\mathbf{D} = \frac{-\sum_{k=1}^{m} c_k u_k}{a_1 + \sum_{k=1}^{m} u_k b_{1k}} \begin{bmatrix} 0 & 0 & \cdots & 0 \\ 0 & 0 & & \vdots \\ \vdots & \vdots & & \vdots \\ b_{11} & b_{12} & \cdots & b_{1m} \end{bmatrix} + \mathbf{C}. \tag{2.20}$$

The matrix **E** is simply of the canonical phase-variable form.

The criterion to be satisfied is that a $\mathbf{v} \in R^m$ is to be found such that the vectors $\mathbf{Dv}, \mathbf{EDv}, \ldots, \mathbf{E}^{(n-1)}\mathbf{Dv}$ are linearly independent. By inspection of (2.20), if **D** is not identically zero, it has nonzero entries only in the bottom row, and there exists a **v** such that **Dv** has the nth component nonzero and all others zero, $\mathbf{E}^2\mathbf{Dv}$ has the $(n-2)$th component nonzero, and so on. Such a set is certainly linearly independent.

There remains the question of whether **D** is identically zero. By inspection of (2.20), if $\mathbf{C} = \mathbf{0}$, then $\mathbf{D} = \mathbf{0}$ and the criterion is not satisfied. Suppose $\mathbf{C} \neq \mathbf{0}$, and for the particular value of **u** under consideration, $\sum_{k=1}^{m} \mathbf{c}_k u_k = 0$. Then $\mathbf{D} = \mathbf{C}$ and the criterion is satisfied. Finally, if $\sum_{k=1}^{m} \mathbf{c}_k u_k \neq 0$, select $\mathbf{v} = \mathbf{u}$ and then

$$\mathbf{Dv} = \mathbf{Du} = \begin{bmatrix} 0 \\ 0 \\ \vdots \\ \sum_{k=1}^{m} c_k u_k \left(1 - \frac{\sum_{k=1}^{m} b_{1k} u_k}{a_1 + \sum_{k=1}^{m} u_k b_{1k}}\right) \end{bmatrix}.$$

This is not identically zero since **A** nonsingular implies $a_1 \neq 0$. Therefore the Lee–Markus criterion is always satisfied if **C** is not identically zero for phase-variable systems.

As observed in the next section for phase-variable systems, condition (a) of the main result may be satisfied even if one or more eigenvalues pass through zero. On the other hand, this condition cannot be satisfied with scalar control ($m = 1$) in a state space of odd dimension n. For then an odd number of eigenvalues must be shifted across the imaginary axis, and at least one of these must pass through zero, since at most $(n-1)/2$ can cross as complex-conjugate pairs. In Section 2.4 it is shown that for $m = 1$, however, the branches of the equilibrium curve are disjoint.

Example 1 *Completely controllable bilinear system.* The second-order system

$$\dot{x}_1 = x_2$$

$$\dot{x}_2 = x_1 - x_2 + u,$$

which is discussed in Section 2.2.1, is completely controllable for unbounded control. For bounded control, however, this system is only locally controllable in some vicinity of the origin. If the process admits an appropriate bilinear mode of control, the system is completely controllable since the Lee–Markus criterion is already satisfied. For example, consider the bilinear system

$$\dot{x}_1 = x_2$$

$$\dot{x}_2 = x_1 - (1 + 4u)x_2 + u,$$

where $|u| \leq 1$ has the eigenvalues

$$\lambda_{1,2}(u) = -(1 + 4u)/2 \pm \tfrac{1}{2}((1 + 4u)^2 - 4)^{1/2}$$

for constant u. For $u = -1$ both are real and positive, crossing the imaginary axis as a complex-conjugate pair. Thus, u^+ and u^- exist and the equilibrium set is connected. The system is of the phase-variable type and $\mathbf{C} \neq \mathbf{0}$, so that the Lee–Markus criterion is satisfied. Thus the new system is completely controllable.

As shown by the next example, however, not all bilinear systems with the capability of transferring system eigenvalues across the imaginary axis are completely controllable. Such is the case even though the system satisfies the Lee–Markus criterion.

Example 2 *A bilinear system without complete controllability.* Consider

$$\dot{x}_1 = 2ux_1 + x_2$$

$$\dot{x}_2 = x_1 + 2ux_2 + u,$$

with eigenvalues $\lambda_1(u) = 2u + 1$, $\lambda_2(u) = 2u - 1$. If the constraint set U is given by $|u| \leq 1$, then the values u^+ and u^- exist. It is easily verified that the Lee–Markus criterion is satisfied for this system. The equilibrium set, however, is not connected, since $m = 1$, and both eigenvalues pass through zero as they cross the imaginary axis in the complex plane. Therefore, the criteria for complete controllability are not all satisfied.

The phase-plane portraits for $u = +1$ and $u = -1$ are shown superimposed in Fig. 2.1. The solid trajectories are for the system with $u = -1$, which has a stable node at $x_1 = -\frac{1}{3}$, $x_2 = -\frac{2}{3}$. The dashed trajectories are for the system with $u = +1$, which has an unstable node at $x_1 = \frac{1}{3}$, $x_2 = -\frac{2}{3}$. The equilibrium set is shown as the three heavy solid curves.

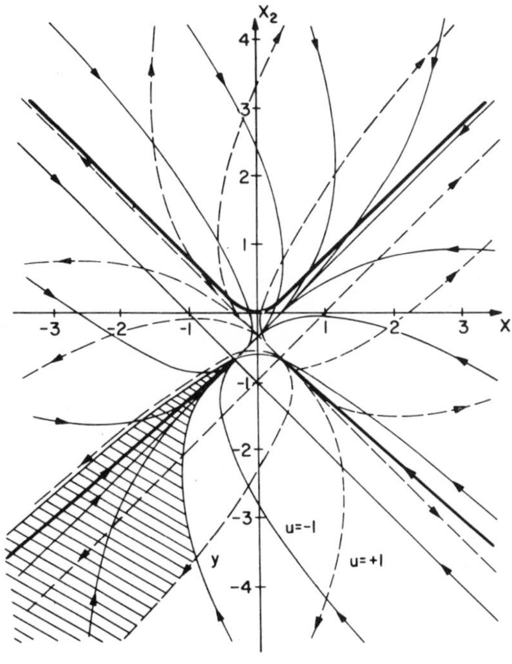

Fig. 2.1. Bilinear phase trajectories for Example 2.

The reachable zones from various initial phases are easily determined on this figure by simply considering the directions of allowable motion at each point, which is the cone between the extremal directions corresponding to $u = +1$ and $u = -1$. It is evident that the system is not completely controllable. For example, the reachable zone from the point y is just the shaded region.

A bilinear system is completely controllable if the conditions given by the main result are satisfied. As one might expect, the conditions are not as simple as the popular conditions for complete controllability of linear systems with unconstrained control or for null controllability with constrained control. For phase-variable systems, however, the sufficient conditions are easy to apply.

In practice, a bilinear mode may be implemented by controlling significant plant parameters in a manner similar to the variable wing geometry of high-performance aircraft. For Example 1, a simple bilinear control made a locally controllable linear system completely controllable. With respect to many systems which are inherently bilinear, however, controllability is further complicated by numerous state constraints. Unfortunately, state constraints can appear in such diversity that meaningful controllability conditions would have to be specified for particular cases. For the classical neutron kinetics of a reactor, it is obvious by inspection of the model that every desired equilibrium state can be attained in finite time. Again such controllability is a consequence of the bilinear control mode, and is analyzed in Chapter IV.

Gershwin and Jacboson [14] further analyze the inherent relationship between controllability and stability for nonlinear systems which are linear in control. Their work is based on the basic concept of controllability and the use of Liapunov functions similar to that done in Section 2.2. In this manner, they derive sufficiency conditions for complete controllability and for uncontrollability based on the existence of Liapunov-like functions.

2.4 EQUILIBRIUM SET OF BILINEAR SYSTEMS

Here, the set of equilibrium points is described for bilinear systems. This description is necessary to understand the connectedness property that is utilized to show complete controllability. For each fixed $\mathbf{u} \in U$, the state

$$\mathbf{x}_e(\mathbf{u}) = -\left(\mathbf{A} + \sum_{k=1}^{m} u_k \mathbf{B}_k\right)^{-1} \mathbf{C}\mathbf{u} \tag{2.21}$$

is the unique equilibrium state of the bilinear system (2.14) if the indicated inverse matrix exists. If the system matrix is singular for a control value $\mathbf{u} = \hat{\mathbf{u}}$, no equilibrium state exists corresponding to $\hat{\mathbf{u}}$ unless $\mathbf{C}\hat{\mathbf{u}}$ happens to lie in the range of $\mathbf{A} + \sum_{k=1}^{m} \mathbf{u}_k \mathbf{B}_k$, in which case an infinite number of states are equilibrium states. It will be assumed here that $\mathbf{C}\mathbf{u}$ is not contained in the range of $\mathbf{A} + \sum_{k=1}^{m} \mathbf{u}_k \mathbf{B}_k$ for any $\mathbf{u} \in U$ such that the latter matrix is singular. This implies in particular that $\mathbf{A}$ is nonsingular, since if $\mathbf{A}$ is singular, $\mathbf{C}\mathbf{0} = \mathbf{0}$ is the range of the singular matrix $\mathbf{A} + \sum_{k=1}^{m} \mathbf{0}\mathbf{B}_k$. It also implies that $\mathbf{C}$ is not identically zero if $\mathbf{A} + \sum_{k=1}^{m} u_k \mathbf{B}_k$ is singular for any $\mathbf{u} \in U$. With this assumption, (2.21) defines a mapping from $U - S$, where S is the subset of U on which the system matrix is singular, onto a subset of R^n, called the equilibrium set.

The equilibrium set is easily described when u is a scalar. Since $\det(\mathbf{A} + u\mathbf{B})$ is a polynomial in u with degree at most n, the equation $\det(\mathbf{A} + u\mathbf{B}) = 0$ has at most n real roots $\{u^i\}$ in U. As u approaches a value $u_f \in U$ for which $\mathbf{A} + u_f\mathbf{B}$ is singular, the curve $\mathbf{x}_e(u)$ tends to infinity asymptotic to the null space of $\mathbf{A} + u_f\mathbf{B}$.

To prove this assertion, let $\{\mathbf{y}_1, \mathbf{y}_2, \ldots, \mathbf{y}_r\}$ be a basis of the null space of $\mathbf{A} + u_f\mathbf{B}$, and let $\{\mathbf{y}_1, \mathbf{y}_2, \ldots, \mathbf{y}_r, \mathbf{y}_{r+1}, \ldots, \mathbf{y}_n\}$ be a basis of R^n. Then one has the unique representation

$$\mathbf{x}_e(u) = \xi_1(u)\mathbf{y}_1 + \xi_2(u)\mathbf{y}_2 + \cdots + \xi_n(u)\mathbf{y}_n \tag{2.22}$$

for every $\{u^i\}$. The equation $(\mathbf{A} + u\mathbf{B})\mathbf{x}_e(u) = -u\mathbf{C}$ must be satisfied, but as $u \to u_f$ the left-hand side has $\xi_i(u)(\mathbf{A} + u\mathbf{B})\mathbf{y}_i \to 0$ for $i = 1, 2, \ldots, r$ unless $|\xi_i(u)| \to \infty$. Since by assumption $u_f\mathbf{C}$ is not contained in the range of $\mathbf{A} + u_f\mathbf{B}$, one or more of the functions $\xi_1(u), \xi_2(u), \ldots, \xi_r(u)$ must tend to infinity as $u \to u_f$. Thus, at each of the control values $\{u^i\}$, the otherwise smooth curve $\mathbf{x}_e(u)$ blows up, and the equilibrium set $\{\mathbf{x}_e(u) \mid u \in U - S\}$ consists of at most $n + 1$ smooth curves having no finite endpoints, with the possible exception of the equilibrium states corresponding to the minimum and maximum values of u in U.

Further, it will now be shown that these curves do not intersect one another. For, if there exists a common point $\mathbf{x}^* = \mathbf{x}_e(u^a) = \mathbf{x}_e(u^b)$, with

$u^a \neq u^b$, then

$$\mathbf{Ax}^* + u^a\mathbf{Bx}^* + u^a\mathbf{C} = \mathbf{Ax}^* + u^b\mathbf{Bx}^* + u^b\mathbf{C} = \mathbf{0}, \qquad (2.23)$$

which implies that $(u^a - u^b)(\mathbf{Bx}^* + \mathbf{C}) = \mathbf{0}$. Thus, from (2.23), $\mathbf{Ax}^* = \mathbf{0}$, and since $\mathbf{x}^*$ is not the zero vector, $\mathbf{A}$ must be singular, contrary to the previous assumption. Therefore, the curves do not intersect one another, but constitute an unconnected set. Of course, if $\{u^i\}$ is empty, the equilibrium set is compact and connected.

Extending these considerations to the case where $m > 1$, suppose that $\mathbf{u}^0 \in U - S$ is a control value for which the mapping (2.21) exists. There is an R^m-neighborhood of $\mathbf{u}^0$ on which (2.21) exists, since $\det(\mathbf{A} + \sum_{k=1}^m u_k{}^0\mathbf{B}_k) \neq 0$ and the determinant of a matrix is a continuous function of the matrix entries. Hence (2.21) defines a continuous mapping of an R^m-neighborhood of $\mathbf{u}^0$ into an R^n-neighborhood of $\mathbf{x}_e(\mathbf{u}^0)$, which geometrically corresponds to an equilibrium surface of dimension $d \leq \min(m, n)$ passing through $\mathbf{x}_e(\mathbf{u}^0)$. As in the scalar case, this surface will tend to infinity as $\mathbf{u}$ approaches a vector value $\mathbf{u}^1$ for which $\mathbf{A} + \sum_{k=1}^m u_k\mathbf{B}_k$ is singular.

If $\det(\mathbf{A} + \sum_{k=1}^m u_k\mathbf{B}_k)$ is thought of as a function of a single component of the control vector, say u_1, with the remaining $m - 1$ components fixed, then

$$P(u_1; u_2{}^0, \ldots, u_m{}^0) = \det\left(\mathbf{A} + u_1\mathbf{B}_1 + \sum_{k=2}^{m} u_k{}^0\mathbf{B}_k\right) \qquad (2.24)$$

is a polynomial of degree at most n in u_1, and has at most n real roots $\{u_1{}^i(u_2{}^0, \ldots, u_m{}^0)\}$ in U. Since the roots of a polynomial vary continuously with the polynomial coefficients, as $u_2, u_3, \ldots, u_m$ range in a small neighborhood about the fixed values $u_2{}^0, u_3{}^0, \ldots, u_m{}^0$, each root $u_1{}^i$ of (2.24) varies continuously and describes an $(m - 1)$-dimensional root surface in U. [This statement needs qualification for points $(u_2{}^0, u_3{}^0, \ldots, u_m{}^0)$ at which the number of real roots of (2.24) changes. Here two root surfaces intersect and the corresponding roots are complex at certain points in every neighborhood of $(u_2{}^0, \ldots, u_m{}^0)$.] In any event, whether the root surfaces intersect or not, they partition U into a number of connected cells, each of which is mapped by (2.21) into a smooth, connected equilibrium surface in state space. Unlike the scalar case, these equilibrium surfaces are not necessarily disjoint, as will now be demonstrated for the phase-variable type of bilinear system.

Consider the bilinear system governed by a single differential equation

$$\frac{d^n x}{dt^n} = \sum_{i=1}^{n} \left(a_i + \sum_{k=1}^{m} u_k b_{ik} \right) \frac{d^{(i-1)}x}{dt^{(i-1)}} + \sum_{k=1}^{m} c_k u_k.$$

With the usual phase variables $x_1 = x, x_2 = \dot{x}, \ldots, x_n = x^{(n-1)}$, the system matrix is singular for control vectors which satisfy

$$\det\left(\mathbf{A} + \sum_{k=1}^{m} u_k \mathbf{B}_k\right) = (-1)^{n-1}\left(a_1 + \sum_{k=1}^{m} u_k b_{1k}\right) = 0. \quad (2.25)$$

For this type of system, it is easily verified that the mapping (2.21) reduces to

$$\mathbf{x}_e(\mathbf{u}) = \frac{-\sum_{k=1}^{m} c_k u_k}{a_1 + \sum_{k=1}^{m} u_k b_{1k}} \begin{bmatrix} 1 \\ 0 \\ \vdots \\ 0 \end{bmatrix}, \quad (2.26)$$

which just expresses the trivial fact that the phase-variable system is at equilibrium only when the dependent variable x is constant and all its derivatives are zero.

The mapping (2.26) fails to exist when (2.25) is satisfied, which can happen for at most a single value of u_1 with given values $u_2, u_3, \ldots, u_m$, due to the linear nature of (2.25). Thus U is partitioned into at most two connected cells by the root surface of (2.25), each of which is mapped onto an interval of the x_1 axis by (2.26). If (2.25) has no roots $\mathbf{u} \in U$, then U is mapped onto a single compact interval of the x_1 axis by (2.26).

Suppose U is partitioned by a root surface S into U^+ and U^- with the denominator of (2.26) positive in U^+ and negative in U^-. Then $\{\mathbf{x}_e(\mathbf{u}) \mid u \in U^+\}$ and $\{\mathbf{x}_e(\mathbf{u}) \mid \mathbf{u} \in U^-\}$ overlap (in fact, each is the entire x_1 axis) if the surface T defined by $\sum_{k=1}^{m} c_k u_k = 0$ intersects S at a nonzero angle in the interior of U. For, in this case, there exists a $\rho > 0$ such that, for any r with $|r| \leq \rho$, the closure of U^+ and the closure of U^- contain sets U_r^+ and U_r^- which intersect S and on which $\sum_{k=1}^{m} c_k u_k = r$. Upon inspection of (2.26), it is clear that

$$\left\{\mathbf{x}_e(\mathbf{u}) \mid \mathbf{u} \in \bigcup_{|r| \leq \rho} U_r^+\right\} \quad \text{and} \quad \left\{\mathbf{x}_e(\mathbf{u}) \mid \mathbf{u} \in \bigcup_{|r| \leq \rho} U_r^-\right\}$$

both cover the entire x_1 axis in R^n.

2.5 REACHABLE ZONE FOR EQUICONTINUOUS CONTROL

Geometrical properties of reachable states are analyzed here for bilinear control systems. Now there is a slight alteration in the admissible class of controls. Piecewise continuous controls are utilized above and usually are assumed for mathematical convenience. Such control may be approximated very closely for many practical control systems. In this section it is more convenient to consider only the class of vector functions that are bounded equicontinuous functions of time [15]. The analysis follows that presented by Mohler and Rink [16].

Again, the bilinear system is described by (2.14) with $\mathbf{u}(t) \in U \subset R^m$, where U is a compact connected set containing the origin, and the admissible controls $\{\mathbf{u}(t)\}$ are equicontinuous on $[0, \infty)$.

2.5.1 Closed Reachable Zone and Local Cone of Tangents

For a given initial state $\mathbf{x}_0$ and a fixed time interval $[0, T]$, consider the set of all trajectories of the bilinear system (2.14), corresponding to all admissible control policies $\mathbf{u}(t)$ over the interval $[0, T]$. The union of these trajectories is the reachable zone from $\mathbf{x}_0$ with elapsed time of at most T, and will be denoted by $R(\mathbf{x}_0, T)$. Since the trajectories of a dynamical system with bounded control effort are extended continuously in time and at a finite rate, $R(\mathbf{x}_0, T)$ is obviously a connected, continuously growing set which shrinks to $\mathbf{x}_0$ as $T \to 0$. The time evolution of this set is described in Section 2.5.2.

First, it is shown that $R(\mathbf{x}_0, T)$ is closed as a subset of R^n for each fixed T, or that every limit point of $R(\mathbf{x}_0, T)$ is reachable from $\mathbf{x}_0$ with $t \leq T$. Suppose that $\mathbf{y}$ is an arbitrary limit point of $R(\mathbf{x}_0, T)$. Then there exists a sequence $\{\mathbf{y}^1, \mathbf{y}^2, \mathbf{y}^3, \ldots\}$ of points in $R(\mathbf{x}_0, T)$ which converges to $\mathbf{y}$. For each $\mathbf{y}^i$ there exists an admissible control policy $\mathbf{u}^i(t)$ over some time interval $[0, t^i] \subset [0, T]$ which steers the system from $\mathbf{x}_0$ to $\mathbf{y}^i$ at time t_i. Since $[0, T]$ is a compact set, a subsequence of $\{u^i(t)\}$ can be chosen for which the terminal times converge to some $t^* \in [0, T]$. But this subsequence belongs to the bounded, equicontinuous family of admissible control policies, so by Arzela's theorem [15] it has a further subsequence which converges to some admissible control policy $\mathbf{u}^*(t)$. There corresponds to this final subsequence of control policies a subsequence of $\{\mathbf{y}^i\}$ which also converges to $\mathbf{y}$. Now, a trajectory of the system of differential equations (2.14) certainly varies continuously as the control policy is varied; hence it follows that $\mathbf{u}^*(t)$ transfers the system from $\mathbf{x}_0$ to $\mathbf{y}$ at time t^*. Therefore $\mathbf{y}$ is contained in $R(\mathbf{x}_0, T)$, and consequently $R(\mathbf{x}_0, T)$ is closed.

The directions of allowable motion from any initial state $\mathbf{x}$ are simply the range of the vector derivative $\dot{\mathbf{x}}$ in (2.14) as a function of $\mathbf{u}$ with domain U. For each $\mathbf{u} \in U$, $\dot{\mathbf{x}}(\mathbf{x}, \mathbf{u})$ is the tangent vector to a trajectory of the bilinear system passing through $\mathbf{x}$. The positive multiples of all such tangents constitute a cone, denoted by $L(\mathbf{x})$, which is described as follows:

The m vectors $(\mathbf{B}_k\mathbf{x} + \mathbf{c}_k)$, $k = 1, 2, \ldots, m$, span a subspace of dimension $d \leq m$ (where $\mathbf{c}_k$ is the kth column of $\mathbf{C}$). Each of these vectors appears in (2.14) multiplied by a scalar u_k with $|u_k| \leq 1$. An additional vector $\mathbf{Ax}$ appears in the linear combination (2.14), but its coefficient is fixed equal to unity.

Clearly $\{\dot{\mathbf{x}}(\mathbf{x}, \mathbf{u}) \mid \mathbf{u} \in U\}$ is a closed set since U is closed. Also $\{\dot{\mathbf{x}}(\mathbf{x}, \mathbf{u}) \mid \mathbf{u} \in U\}$ is convex, since if $\dot{\mathbf{x}}(\mathbf{x}, \mathbf{u}^1)$ and $\dot{\mathbf{x}}(\mathbf{x}, \mathbf{u}^2)$ are any two tangents and $\alpha + \beta = 1$, then

$$\alpha\dot{\mathbf{x}}(\mathbf{x}, \mathbf{u}^1) + \beta\dot{\mathbf{x}}(\mathbf{x}, \mathbf{u}^2) = \mathbf{Ax} + \sum_{k=1}^{m} (\alpha u_k{}^1 + \beta u_k{}^2)(\mathbf{B}_k\mathbf{x} + \mathbf{c}_k) \tag{2.27}$$

is in $\{\dot{\mathbf{x}}(\mathbf{x}, \mathbf{u}) \mid \mathbf{u} \in U\}$ since U is convex.

Now two cases are possible. For a fixed $\mathbf{x}$, either $-\mathbf{Ax}$ can be represented as

$$-\mathbf{Ax} = \sum_{k=1}^{m} u_k{}^e(\mathbf{B}_k\mathbf{x} + \mathbf{c}_k) \tag{2.28}$$

for some $\mathbf{u}^e \in U$, or else no such representation exists. In the former case, $\mathbf{x}$ is in the equilibrium set of the system, and $\{\dot{\mathbf{x}}(\mathbf{x}, \mathbf{u}) \mid \mathbf{u} \in U\}$ is a closed convex set containing the origin in R^n. Also, (2.14) can then be written in the form

$$\dot{\mathbf{x}}(\mathbf{x}, \mathbf{u}) = \sum_{k=1}^{m} (u_k - u_k{}^e)(\mathbf{B}_k\mathbf{x} + \mathbf{c}_k). \tag{2.29}$$

If representation (2.28) exists for some $\mathbf{u}^e$ in the interior of U, then each coefficient on the right-hand side of (2.29) can take on either positive or negative values, and it follows that $L(\mathbf{x})$ is just the linear subspace spanned by $(\mathbf{B}_k\mathbf{x} + \mathbf{c}_k)$, $k = 1, 2, \ldots, m$. However, if the only $\mathbf{u}^e$ which satisfies (2.28) is on the boundary of U, then some coefficients in (2.29) are constrained to be only positive or only negative.

If no representation of the form (2.28) exists, $\{\dot{\mathbf{x}}(\mathbf{x}, \mathbf{u}) \mid \mathbf{u} \in U\}$ is a closed convex set which does not contain the origin; hence it is contained in the interior of some half-space whose bounding hyperplane passes through the origin. In this case $\mathbf{x}$ is not an equilibrium point of the system, and all rays of $L(\mathbf{x})$ point into the interior of the half-space above.

Now, for each $\mathbf{x}$, $L(\mathbf{x})$ is a closed convex cone of one of the above-mentioned types, with vertex at the origin in R^n. It is necessary to describe the geometry and to define a notion of continuity for $L(\mathbf{x})$. First, a metric must be defined for rays and cones.

Consider cones $L(\mathbf{x}^1)$ and $L(\mathbf{x}^2)$ corresponding to different states $\mathbf{x}^1$ and $\mathbf{x}^2$. If $L(\mathbf{x}^2)$ contains a ray $\mathbf{r}^1$ corresponding to the tangent vector $\dot{\mathbf{x}}(\mathbf{x}^1, \mathbf{u}^1)$, and $L(\mathbf{x}^2)$ contains a ray $\mathbf{r}^2$ corresponding to the tangent vector $\dot{\mathbf{x}}(\mathbf{x}^2, \mathbf{u}^2)$, the distance between $\mathbf{r}^1$ and $\mathbf{r}^2$ is defined as the usual vector space distance between the unit generators of these rays:

$$d(\mathbf{r}^1, \mathbf{r}^2) = \left\| \frac{\dot{\mathbf{x}}(\mathbf{x}^1, \mathbf{u}^1)}{\|\dot{\mathbf{x}}(\mathbf{x}^1, \mathbf{u}^1)\|} - \frac{\dot{\mathbf{x}}(\mathbf{x}^2, \mathbf{u}^2)}{\|\dot{\mathbf{x}}(\mathbf{x}^2, \mathbf{u}^2)\|} \right\|.$$

The distance from $\mathbf{r}^2$ to the cone $L(\mathbf{x}^1)$ is then defined as the distance from $\mathbf{r}^2$ to the "nearest" ray in $L(\mathbf{x}^1)$, or

$$d\big(L(\mathbf{x}^1), \mathbf{r}^2\big) = \min_{\mathbf{r}^1 \in L(\mathbf{x}^1)} d(\mathbf{r}^1, \mathbf{r}^2).$$

This distance exists since $L(\mathbf{x}^1)$ is a closed cone. Finally, the distance between $L(\mathbf{x}^1)$ and $L(\mathbf{x}^2)$ could be taken as the distance from the "furthest" ray in $L(\mathbf{x}^2)$ to $L(\mathbf{x}^1)$. To make the definition symmetric, however, let

$$d\big(L(\mathbf{x}^1), L(\mathbf{x}^2)\big) = \max_{\mathbf{r}^2 \in L(\mathbf{x}^2)} \min_{\mathbf{r}^1 \in L(\mathbf{x}^1)} \left[\frac{d\{\mathbf{r}^1, \mathbf{r}^2)}{2}\right] + \max_{\mathbf{r}^1 \in L(\mathbf{x}^1)} \min_{\mathbf{r}^2 \in L(\mathbf{x}^2)} \left[\frac{d(\mathbf{r}^1, \mathbf{r}^2)}{2}\right].$$

The definition of continuity for $L(\mathbf{x})$ is then as follows: $L(\mathbf{x})$ is said to be continuous at $\mathbf{x}^0$ if and only if for every $\varepsilon > 0$ there exists a $\delta(\varepsilon) > 0$ such that

$$d\big(L(\mathbf{x}), L(\mathbf{x}^0)\big) < \varepsilon$$

for every $\mathbf{x}$ such that

$$\|\mathbf{x} - \mathbf{x}^0\| < \delta.$$

The equivalent definition in terms of convergent sequences is as follows: $L(\mathbf{x})$ is continuous at $\mathbf{x}$ if and only if for every sequence $\{\mathbf{x}^i\}_{i=1}^{\infty}$ converging to $\mathbf{x}^0$, the sequence of numbers

$$\{d(L(\mathbf{x}^i), L(\mathbf{x}^0))\}_{i=1}^{\infty}$$

converges to zero. This form of the definition is used in the following discussion.

First, suppose that $\mathbf{x}^0$ is not in the equilibrium set of the bilinear system (2.14). The equilibrium set is closed, so the distance from $\mathbf{x}^0$ to the equilibrium set is positive. Therefore, if $\{\mathbf{x}^i\}_{i=1}^{\infty}$ is a sequence of points converging to $\mathbf{x}^0$, there exists an integer N such that $\{\mathbf{x}^i\}_{i=N}^{\infty}$ contains no equilibrium points. Then, for every $\mathbf{u} \in U$, the sequence of unit generators

$$\{\dot{\mathbf{x}}(\mathbf{x}^i, \mathbf{u})/\|\dot{\mathbf{x}}(\mathbf{x}^i, \mathbf{u})\|\}_{i=N}^{\infty}$$

is defined and converges to $\dot{\mathbf{x}}(\mathbf{x}^0, \mathbf{u})/\|\dot{\mathbf{x}}(\mathbf{x}^0, \mathbf{u})\|$. Thus the unit generators of the cones in the sequence $\{L(\mathbf{x}^i)\}_{i=N}^{\infty}$ converge to the unit generators of $L(\mathbf{x}^0)$ with a one-to-one correspondence, and it follows that

$$\{d(L(\mathbf{x}^i), L(\mathbf{x}^0))\}_{i=1}^{\infty}$$

converges to zero. Therefore $L(\mathbf{x})$ is continuous at every $\mathbf{x}^0$ not in the equilibrium set.

However, $L(\mathbf{x})$ need not be continuous at points in the equilibrium set. In particular, it will be shown that if $\mathbf{x}^0$ is an equilibrium point for which the only equilibrium control value is on the boundary of U, then $L(\mathbf{x})$ is not continuous at $\mathbf{x}^0$. For, in this case, every neighborhood of $\mathbf{x}^0$ contains an equilibrium point $\dot{\mathbf{x}}$ with equilibrium control value $\mathbf{u}_e(\mathbf{x})$ in the interior of U, so that $L(\mathbf{x})$ contains the ray $\mathbf{r} = \{-\alpha\mathbf{A}\mathbf{x}^0 \mid 0 \le \alpha < \infty\}$. Consider a sequence $\{\mathbf{x}^i\}_{i=1}^{\infty}$ of such points which converges to $\mathbf{x}^0$. The corresponding sequence of rays $\{\mathbf{r}^i\}_{i=1}^{\infty}$ obviously converges to $\mathbf{r} = \{-\alpha\mathbf{A}\mathbf{x}^0 \mid 0 \le \alpha < \infty\}$. This ray, however, is not contained in $L(\mathbf{x}^0)$, since if for some $\mathbf{u} \in U$

$$\dot{\mathbf{x}}(\mathbf{x}^0, \mathbf{u}) = \mathbf{A}\mathbf{x}^0 + \sum_{k=1}^{m} u_k(\mathbf{B}_k\mathbf{x}^0 + \mathbf{c}_k) = -\beta\mathbf{A}\mathbf{x}^0,$$

with $\beta > 0$, then

$$\sum_{k=1}^{m} \frac{u_k}{(1 + \beta)} (\mathbf{B}_k\mathbf{x}^0 + \mathbf{c}_k) = -\mathbf{A}\mathbf{x}^0.$$

This implies that $\mathbf{u}/(1 + \beta)$ is an equilibrium control value in the interior of U, which contradicts the hypothesis. Therefore $L(\mathbf{x})$ is not continuous at $\mathbf{x}^0$.

2.5.2 Time Evolution of the Reachable Zone

Let $\mathbf{x}^0$ be an initial state which is not an equilibrium state. The equilibrium set of the system (2.14) was seen to be a closed set; therefore $\mathbf{x}^0$ is at a positive distance from the equilibrium set, and $R(\mathbf{x}_0, T)$ does not intersect the equilibrium set for T sufficiently small, $T > 0$. A geometric description of $R(\mathbf{x}_0, T)$ is given in this section for all T prior to the first time T_1 such that $R(\mathbf{x}_0, T_1)$ contains an equilibrium point. (T_1 may not exist, of course, if the system is not completely controllable.)

In the previous section, $L(\mathbf{x})$ was seen to be a cone with vertex at the origin for each $\mathbf{x} \in R^n$. Now imagine that for each $\mathbf{x} \in R^n$, $L(\mathbf{x})$ is translated in space so that its vertex is at $\mathbf{x}$. When this is done, each trajectory of the bilinear system passing through $\mathbf{x}$ is tangent to a ray of the cone $L(\mathbf{x})$ which is "attached" to the point $\mathbf{x}$. In the development to follow, this interpretation is assumed, so that R^n is covered by the family of attached local cones.

First, consider the nature of the attached local cones $L(\mathbf{x})$ for points $\mathbf{x}$ in a small neighborhood of the fixed initial point $\mathbf{x}_0$. Since $\mathbf{x}_0$ is not an equilibrium state, all rays of $L(\mathbf{x}_0)$ point into the interior of a half-space H bounded by a hyperplane P passing through $\mathbf{x}_0$; and since $L(\mathbf{x})$ is continuous at $\mathbf{x}_0$, the same is true for all rays of $L(\mathbf{x})$ for every $\mathbf{x}$ in some sufficiently small neighborhood of $\mathbf{x}_0$. It follows from this that for T sufficiently small, $R(\mathbf{x}_0, T)$ is contained in the half-space H. For, if a local trajectory from $\mathbf{x}_0$ crosses P in every neighborhood of $\mathbf{x}_0$, a tangent vector must point into the exterior of H at some point $\mathbf{x}$ in every neighborhood of $\mathbf{x}_0$, which is a contradiction. Therefore the boundary of $R(\mathbf{x}_0, T)$ contains $\mathbf{x}_0$ for T sufficiently small, $T > 0$, and the system is not locally controllable at the nonequilibrium state $\mathbf{x}_0$.

The evolution of the boundary of $R(\mathbf{x}_0, T)$ as a function of T can now be described by considering, for each fixed T, the local cone of tangents attached to each boundary point of $R(\mathbf{x}_0, T)$. If $\mathbf{y}$ is such a boundary point, and if no ray of the attached cone $L(\mathbf{y})$ points into the exterior of $R(\mathbf{x}_0, T)$, then the boundary at $\mathbf{y}$ is not being extended as a function of T. However, if some ray of $L(\mathbf{y})$ enters the exterior of $R(\mathbf{x}_0, T)$ at $\mathbf{y}$, then since $\mathbf{y}$ is reachable from $\mathbf{x}_0$ in time at most T, the boundary at $\mathbf{y}$ will certainly be extended as T increases.

Thus the boundary of $R(\mathbf{x}_0, T)$ consists of two essentially different parts. One part, which will be called the lateral boundary, contains $\mathbf{x}_0$ for T small, and has the boundary rays of the attached local cone $L(\mathbf{x}_0)$ as its tangents as $T \to 0$. This portion of the boundary is stationary in time in the sense that if $\mathbf{y}$ is a lateral boundary point of $R(\mathbf{x}_0, T)$, it is also a lateral boundary point of $R(\mathbf{x}_0, T^*)$ for some $T^* > T$. The other part of the boundary will be called the front, by analogy with a propagating wave front. This is where $R(\mathbf{x}_0, T)$ is being extended as T increases, and this portion is nonstationary in the sense that if $\mathbf{y}$ is a front boundary point of $R(\mathbf{x}_0, T)$, it is not a front boundary point of $R(\mathbf{x}_0, T^*)$ for $T^* \neq T$. These concepts are shown for the case $n = 2$, $m = 1$ in Fig. 2.2.

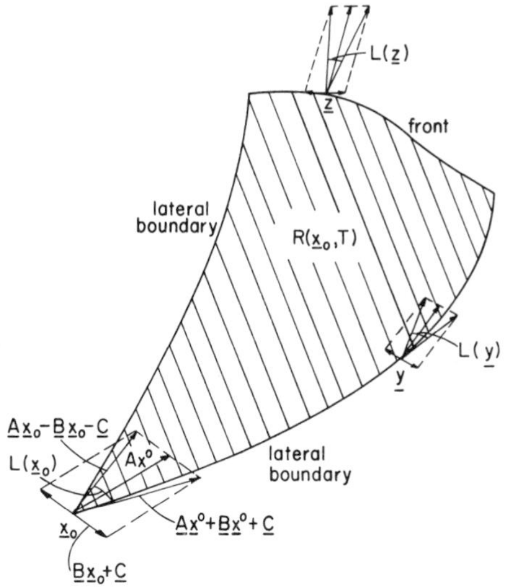

Fig. 2.2. Qualitative features of reachable set ($n = 2$, $m = 1$).

Now suppose that $\mathbf{y}$ is a lateral boundary point of $R(\mathbf{x}_0, T)$ for some fixed T, so that a stationary boundary surface of $R(\mathbf{x}_0, T)$ contains $\mathbf{y}$. Then it is necessarily true that at least one ray of the attached cone $L(\mathbf{y})$ is tangent to this boundary surface at $\mathbf{y}$. For, if all rays of $L(\mathbf{y})$ point into the interior of $R(\mathbf{x}_0, T)$ at $\mathbf{y}$, the same is true for all rays of $L(\mathbf{z})$ for every point $\mathbf{z}$ in a sufficiently small neighborhood of $\mathbf{y}$—again due to the continuity of $L(\mathbf{x})$ at the nonequilibrium point $\mathbf{y}$. Any trajectory from $\mathbf{x}_0$ to $\mathbf{y}$ must approach $\mathbf{y}$ either from the interior of $R(\mathbf{x}_0, T)$ or along the boundary of $R(\mathbf{x}_0, T)$, and these directions of motion are not allowable if all tangents point into the interior of $R(\mathbf{x}_0, T)$, which contradicts the result that $R(\mathbf{x}_0, T)$ is closed and does contain $\mathbf{y}$.

Consequently, for every lateral boundary point $\mathbf{y}$ or $R(\mathbf{x}_0, T)$, the attached cone $L(\mathbf{y})$ has at least one ray tangent to the stationary boundary surface containing $\mathbf{y}$, but has no rays entering the exterior of $R(\mathbf{x}_0, T)$ at $\mathbf{y}$. Therefore the lateral boundary of $R(\mathbf{x}_0, T)$ is just the envelope of its attached cones, starting with $L(\mathbf{x}_0)$.

This concludes the discussion of $R(\mathbf{x}_0, T)$ for $T < T_1$. Next, consider the behavior of $R(\mathbf{x}_0, T)$ as $T \to \infty$. Two cases are possible. Either

(i) the front of $R(\mathbf{x}_0, T)$ goes to infinity in R^n as $T \to \infty$, or
(ii) there exists a compact subset of R^n in which $R(\mathbf{x}_0, T)$ continues to evolve as $T \to \infty$.

In case (i) the reachable zone

$$R(\mathbf{x}_0) = \bigcup_{T=0}^{\infty} R(\mathbf{x}_0, T)$$

is closed since any finite boundary point is a lateral boundary point of $R(\mathbf{x}_0, T)$ for some T, and hence belongs to $R(\mathbf{x}_0)$.

In case (ii), however, $R(\mathbf{x}_0)$ need not be closed since there may exist boundary points of $R(\mathbf{x}_0)$ which are approached by $R(\mathbf{x}_0, T)$ as $T \to \infty$, but are not reachable in finite time. In this case the boundary of $R(\mathbf{x}_0)$ will consist of two parts: the closed boundary, or set of all boundary points which are lateral boundary points of $R(\mathbf{x}_0, T)$ for some T; and the open boundary, or set of all boundary points which are not reachable in finite time.

If the system is locally controllable on the boundary of $R(\mathbf{x}_0)$, then it cannot contain any equilibrium points with equilibrium control values in the interior of U. Portions of the boundary of $R(\mathbf{x}_0)$, however, may be comprised of images of surface elements of U under the equilibrium set mapping:

$$\dot{\mathbf{x}}_e(\mathbf{u}) = -\left(\mathbf{A} + \sum_{k-1}^{m} u_k \mathbf{B}_k\right)^{-1} \mathbf{C}\mathbf{u}.$$

Here the local cone of tangents is discontinuous. The local cone of tangents is continuous on the remaining, nonequilibrium portions of the boundary of $R(\mathbf{x}_0)$, and it follows from the previous discussion that the nonequilibrium closed-boundary portion is just the envelope of its attached local cones.

A somewhat similar conclusion can be arrived at for the nonequilibrium open-boundary portion. If $\mathbf{y}$ is an open-boundary point of $R(\mathbf{x}_0)$, one must consider not the attached local cone $L(\mathbf{y})$, but rather the "inverse" local one $-L(\mathbf{y})$ attached to $\mathbf{y}$. For, even though a ray of $L(\mathbf{y})$ may enter the exterior of $R(\mathbf{x}_0)$ at $\mathbf{y}$, it does not follow that a trajectory of the system will leave $R(\mathbf{x}_0)$ since $\mathbf{y}$ itself is not reachable. However, if the negative of a ray of $L(\mathbf{y})$ points into the interior of $R(\mathbf{x}_0)$ at $\mathbf{y}$, then a trajectory from $\mathbf{x}_0$ will pass through $\mathbf{y}$ tangent to this ray, contrary to the hypothesis that $\mathbf{y}$ is not reachable from $\mathbf{x}_0$. Thus the criterion that $\mathbf{y}$ be a nonequilibrium open-boundary point of $R(\mathbf{x}_0)$ is that no rays of the attached inverse cone $-L(\mathbf{y})$ enter the interior of $R(\mathbf{x}_0)$ at $\mathbf{y}$. Also, by an application of the assumed continuity of $L(\mathbf{x})$ at $\mathbf{y}$ similar to that given previously for lateral boundary points, it follows that at least one ray of $-L(\mathbf{y})$ must be tangent to the open-boundary surface of $R(\mathbf{x}_0)$ at $\mathbf{y}$ for $\mathbf{y}$ to be approachable at $\mathbf{x}_0$. Therefore the nonequilibrium open boundary of $R(\mathbf{x}_0)$ is just the envelope of its attached local inverse cones.

2.5.3 Summary

The boundary of $R(\mathbf{x}_0)$ has the following (possible) elements: a closed portion consisting of the envelope of its attached local cones, and an open portion consisting of the envelope of its attached local inverse cones, separated by equilibrium points that correspond to equilibrium control values on the surface of U, where the local cone is discontinuous. An example of this geometry is furnished by the phase-plane portrait of Fig. 2.1 for Example 2 in Section 2.3.

Here the reachable zone $R(\mathbf{y})$ from the initial point $\mathbf{y}$ has as its boundary the closed-boundary curves Γ_1 and Γ_2 and the open-boundary curve Γ_3. Curves Γ_2 and Γ_3 are separated by the equilibrium point $x_1 = -\frac{1}{3}$, $x_2 = -\frac{2}{3}$, which corresponds to the extreme control value $u = -1$.

All the considerations of $R(\mathbf{x}_0, T)$ and $R(\mathbf{x}_0)$ for an initial state $\mathbf{x}_0$ presented in this section apply equally well to the incident zones $I(\mathbf{x}_f, T)$ and $I(\mathbf{x}_f)$ to a specified terminal point $\mathbf{x}_f$. For, the incident zone to $\mathbf{x}_f$ is just the reachable zone from $\mathbf{x}_f$ for the system with the sense of time reversed. To find $I(\mathbf{x}_f)$ for the system (2.14) it suffices to find $R(\mathbf{x}_f)$ for the system

$$\dot{\mathbf{x}} = -\mathbf{A}\mathbf{x} - \left(\sum_{k=1}^{m} u_k \mathbf{B}_k\right)\mathbf{x} - \mathbf{C}\mathbf{u}. \tag{2.30}$$

This is so because reversing the sense of time merely reverses the direction of motion along a trajectory, and does not alter the geometry of that trajectory. Each trajectory of (2.14) which reachs $\mathbf{x}_f$ at time $t = 0$, with a control policy $\mathbf{u}(t)$ defined on a time interval $[-\tau, 0]$, is identical with the trajectory of (2.30) which leaves $\mathbf{x}_f$ at time $t = 0$ and has the control policy $\mathbf{u}(-t)$ defined on the time interval $[0, \tau]$.

2.6 MISCELLANEOUS CONTROLLABILITY PROPERTIES

While multiplicative control by its variable structure is a valuable asset to controllability, the need for additive control is obvious from the previous work. This point is shown vividly by the simple first-order system

$$\dot{x} = ax + bux + cu,$$

where $|u| \leq 1$. It is immediately obvious that this system is completely controllable if $a = 0$, $b = 0$, and $c \neq 0$. By considering the other combinations of coefficients it is just as easy to show that this is a necessary condition as well.

On the other hand, similar calculations readily show that the system with two independent controls such that

$$\dot{x} = ax + bux + cv,$$

where $|u| \leq 1$, $|v| \leq 1$, is completely controllable if and only if $|b| \geq |a|$ and $c \neq 0$.

There are numerous necessary conditions that can be derived to determine the controllability of bilinear systems. Some of these are presented here with the proofs left as exercises for the reader.

Necessary Conditions for Complete Controllability

(1) For the bilinear system (2.14) with scalar control to be completely controllable, it is necessary that $\mathbf{A}$ has distinct nonzero eigenvalues and that $\mathbf{A}$ and $\mathbf{B}$ have no common eigenvectors.

(2) Define the following sets:

$$\begin{aligned} V_A &= \{\mathbf{x} \mid \mathbf{A}\mathbf{x} = 0\} \\ V_{B_1} &= \{\mathbf{x} \mid \mathbf{B}_1\mathbf{x} = -\mathbf{c}_1\} \\ &\vdots \\ V_{B_m} &= \{\mathbf{x} \mid \mathbf{B}_m\mathbf{x} = -\mathbf{c}_m\}, \end{aligned}$$

where $\mathbf{C} = [\mathbf{c}_1 \quad \mathbf{c}_2 \quad \cdots \quad \mathbf{c}_m]$. Then for the bilinear system to be completely controllable it is necessary that the intersection of these sets is the null set, or

$$V_A \cap V_{B_1} \cap \cdots \cap V_{B_m} = \phi.$$

(3) A necessary condition that the bilinear system (2.14) with scalar control be completely controllable is that

$$\begin{aligned} &\operatorname{rank}[(\mathbf{B}\mathbf{x}_e + \mathbf{c}), (\mathbf{A} + \mathbf{B}u_e)(\mathbf{B}\mathbf{x}_e + \mathbf{c}), \mathbf{B}(\mathbf{B}\mathbf{x}_e + \mathbf{c}), \\ &\qquad \ldots, (\mathbf{A} + \mathbf{B}u_e)^{n-1}(\mathbf{B}\mathbf{x}_e + \mathbf{c}), \ldots, \mathbf{B}^{n-1}(\mathbf{B}\mathbf{x}_e + \mathbf{c})] = n, \end{aligned}$$

at every equilibrium point $\mathbf{x}_e$ with the required equilibrium control u_e. It is relatively easy to generalize this condition for vector control with only a major complication in the algebra.

Example

$$\dot{\mathbf{x}} = \mathbf{A}\mathbf{x} + \mathbf{B}u\mathbf{x} + \mathbf{c}u,$$

where

$$\mathbf{A} = \begin{bmatrix} 1 & 0 \\ 0 & 0 \end{bmatrix}, \quad \mathbf{B} = \begin{bmatrix} 1 & 0 \\ 0 & 2 \end{bmatrix}, \quad \mathbf{c} = \begin{bmatrix} 1 \\ 1 \end{bmatrix}.$$

For condition (2)

$$V_A = \left\{\mathbf{x} \mid \mathbf{x} = \begin{bmatrix} 0 \\ x_2 \end{bmatrix}\right\}$$

$$V_B = \left\{\mathbf{x} \mid \mathbf{x} = \begin{bmatrix} -1 \\ -\frac{1}{2} \end{bmatrix}\right\}.$$

Hence, $V_A \cap V_B = \phi$ as required.

For condition (3), however, at

$$\mathbf{x}_e = \begin{bmatrix} 0 \\ -\frac{1}{2} \end{bmatrix}, \qquad u_e = 0,$$

the rank test yields

$$\operatorname{rank}\begin{bmatrix} 1 & 1 & 1 \\ 0 & 0 & 0 \end{bmatrix} = 1 \neq n.$$

Consequently, the system is not completely controllable despite the satisfaction of condition (2).

Note that if the $\mathbf{B}_k$ matrices vanish, condition (3) reduces to the necessary and sufficient condition so familiar for linear systems.

2.7 STABILITY ANALYSIS

Since the bilinear system with $\mathbf{u}(t)$ specified a priori is a linear though generally time-variant system, certain techniques for linear time-variant systems may be useful. D'Angelo [17] provides a good discussion of linear time-variant systems analysis. Frequency domain stability analysis, for example, may be applied to certain bilinear stability problems as special cases of a more general problem.

Consider the bilinear system with independent scalar additive and multiplicative control (output times control) so that

$$d\mathbf{x}/dt = \mathbf{A}\mathbf{x} + \mathbf{c}[y(t)u(t) + v(t)]$$

and

$$y(t) = \mathbf{d}^{\mathrm{T}}\mathbf{x}(t),$$

where $0 \leq u_2(t) \leq \beta$†. It is assumed that the time-invariant linear portion of the system is asymptotically stable with state-transition matrix $\mathbf{\Phi}(t)$. Also it is assumed that $w(t) = \mathbf{d}^T\mathbf{\Phi}(t - t_0)\mathbf{x}_0$ and $v(t)$ are absolutely square integrable.

The bilinear feedback system by definition is L_2-stable if there exists a positive constant ρ such that

$$\left[\int_0^\infty |y(t)|^2\,dt\right]^{1/2} \leq \rho\left[\int_0^\infty |v(t) + w(t)|^2\,dt\right]^{1/2} + \left[\int_0^\infty |w(t)|^2\,dt\right]^{1/2}$$

for every admissible $w(t)$ and $v(t)$.

Then, from the circle criterion [17], the feedback system is L_2-stable if $\mathrm{Re}[G(j\omega)] > -1/\beta$ for all real ω, where $G(j\omega)$ is the frequency response function of the linear feed-forward portion of the system.

Example Obtain the bounds on $u(t)$ so that the system is stable with

$$\begin{bmatrix}\dot{x}_1\\ \dot{x}_2\end{bmatrix} = \begin{bmatrix}0 & 1\\ 0 & -2\end{bmatrix}\begin{bmatrix}x_1\\ x_2\end{bmatrix} + \begin{bmatrix}0 & 0\\ 1 & 0\end{bmatrix}\begin{bmatrix}x_1\\ x_2\end{bmatrix}u(t) + \begin{bmatrix}0\\ 1\end{bmatrix}v(t)$$

and

$$y = [1 \quad 0]\begin{bmatrix}x_1\\ x_2\end{bmatrix}.$$

It is left as an exercise to obtain the Nyquist plot $G(j\omega)$ and show that $\mathrm{Re}[G(j\omega)] > -\frac{1}{4}$ so that $0 \leq u(t) < 4$.

As for other techniques derived from the work of Popov, and before that from Liapunov, the range of stability is quite conservative.

2.8 CONCLUSIONS

It is shown that bilinear systems as a result of their adaptive structure can be designed to be completely controllable with bounded control. This is a property seldom found in linear control systems, or for that matter in physical problems, but nevertheless it is frequently a most

† A nonzero lower limit can be handled easily by a simple modification to the system.

convenient and powerful property used in the analysis of practical problems. For design of a controllable pursuit system a bilinear mode of control might be added to the usual linear controller. In Chapter III it is shown that such a design can exhibit better performance than that of linear systems.

Apparently, there is some relationship between stability and controllability. This aspect has been studied here. Certainly this relationship has been quite useful; however, it has not been fully exploited here.

Some special attention has been given to bilinear systems with independent additive and multiplicative controls in this chapter. At this point, the reader might question its significance. In Chapters IV and V, however, most of the nuclear reactor processes and physiological processes are described by independent multiplicative and additive controls. It would seem that such bilinear systems are particularly relevant to nature's scheme.

Exercises

2.1 In the definition of complete controllability, can the finite terminal state $\mathbf{x}_f$ be replaced by the origin $\mathbf{0}$ for bilinear systems as it can for linear systems? Why?

2.2 Show that a second-order linear system with imaginary eigenvalues is completely controllable with bounded control just as it is with unbounded control.

2.3 Does complete controllability (local controllability) of bilinear systems with unbounded control imply that the same systems would be completely controllable (locally controllable) with bounded control? Why?

2.4 Analyze controllability of

(a)
$$\dot{x}_1 = x_2$$
$$\dot{x}_2 = 3x_1 + x_2 + ux_1,$$

(b)
$$\dot{x}_1 = x_2$$
$$\dot{x}_2 = 3x_1 + x_2 + u_1x_1 + u_2.$$

2.5 Why might it be said that the controlled use of instability may be utilized to make bilinear systems more controllable than linear systems?

2.6 Describe the equilibrium sets and the reachable zones for the processes given with Problem 2.4.

2.7 Discuss second-order bilinear system synthesis with respect to controllability by superposition of state portraits for linear systems.

2.8 Show that variable-wing geometry for a high-performance aircraft results in a dynamical model that is approximately bilinear. Why does such parametric aircraft control result in significantly better controllability and performance than that with but only the standard additive control?

2.9 Explain how a bilinear system can be completely controllable even if a system eigenvalue passes through the origin as $u(t)$ ranges over the admissible set.

2.10 What is the minimum number of equilibrium points for the nth-order bilinear system with scalar control? Explain.

2.11 Prove that a necessary condition for the nth-order bilinear system with scalar control to be completely controllable is that

$$\operatorname{rank}[(\mathbf{B}\mathbf{x}_e + \mathbf{c}), (\mathbf{A} + \mathbf{B}u_e)(\mathbf{B}\mathbf{x}_e + \mathbf{c}), \mathbf{B}(\mathbf{B}\mathbf{x}_e + \mathbf{c}), \ldots, (\mathbf{A} + \mathbf{B}u_e)^{n-1}(\mathbf{B}\mathbf{x}_e + \mathbf{c}), \ldots, \mathbf{B}^{n-1}(\mathbf{B}\mathbf{x}_e + \mathbf{c})] = n,$$

at every equilibrium state $\mathbf{x}_e$ with the required equilibrium control u_e.

2.12 Explain how an additive negative control force would enhance controllability of a braking car.

REFERENCES

1. Caratheodory, C., "Calculus of Variations and Partial Differential Equations of the First Order." Holden-Day, San Francisco, California, 1967. (Translation by J. L. Brandstatter of German ed., published by Teubner, Berlin, 1935.)
2. Hsu, J. C., and Meyer, A. V., "Modern Control Principles and Application." McGraw-Hill, New York, 1968.
3. Athans, M., and Falb, P. L., "Optimal Control." McGraw-Hill, New York, 1965.
4. Lee, E. B., and Markus, L., "Foundations of Optimal Control Theory." Wiley, New York, 1967.

5. Kalman, R. E., Ho, Y. C., and Narendra, K. S., Controllability of linear dynamical systems. *Contrib. Differential Equations* **1**, 189–213 (1963).
6. Kalman, R. E., Ho, Y. C., and Narendra, K. S., Mathematical description of linear dynamical systems. *SIAM J. Control* **1**, 152–192 (1963).
7. Kučera, J., On accessibility of bilinear systems. *Czechoslovak Math. J.* **20**, 160–168 (1970).
8. Chevalley, C., "Theory of Lie Groups I." Princeton Univ. Press, Princeton, New Jersey, 1946.
9. Kučera, J., Solution in large of control problem $\dot{\mathbf{x}} = [\mathbf{A}(1 - u) + \mathbf{B}u]\mathbf{x}$. *Czechoslovak Math. J.* **16**, 600–622 (1966).
10. Kalman, R. E., Falb, P. L., and Arbib, M. A., "Topics in Mathematical System Theory." McGraw-Hill, New York, 1969.
11. Hermes, H., Controllability and the singular problem. *J. SIAM Control* **2**, 241–260 (1965).
12. Rink, R. E., and Mohler, R. R., Completely controllable bilinear systems. *SIAM J. Control* **6**, 477–486 (1968).
13. Lee, E. B., and Markus, L., Optimal control for nonlinear processes. *Arch. Rational Mech. Anal.* **8**, 36–58 (1961).
14. Gershwin, S. B., and Jacobson, D. H., A controllability theory for nonlinear systems. *IEEE Trans. Automat. Control* **AC-16**, 37–46 (1971).
15. Simmons, G. F., "Introduction to Topology and Modern Analysis." McGraw-Hill, New York, 1963.
16. Mohler, R. R., and Rink, R. E., Reachable zones for equicontinuous bilinear control processes. *Internat. J. Control* **14**, 331–339 (1971).
17. D'Angelo, H., "Linear Time-Varying Systems: Analysis and Synthesis." Allyn & Bacon, Rockleigh, New Jersey, 1970.

III

Optimal Control

The ability to control a system to various states has been studied in Chapter II. There it is shown that bilinear systems, as a consequence of their variable structure, can be more controllable than linear systems. Not only may bilinear systems be controlled over a large region of state space, but it is shown here that they also may attain such controllability with a high degree of performance. It is no accident that these variable-structure systems are so prevalent in nature, or that they are so useful in control system design or policy formulation. Man and nature about him strive for some form of optimality, be it quality of life or probability of kill.

3.1 BACKGROUND

A mathematical cost function or performance index must be formed in order to quantify goodness of control. It is mathematically expedient and yet meaningful to define a cost functional or performance index by

$$J = \int_{t_0}^{t_f} f_0(\mathbf{x}, \mathbf{u})\, dt, \tag{3.1}$$

where $f_0(\mathbf{x}, \mathbf{u})$ and $\partial f_0(\mathbf{x}, \mathbf{u})/\partial \mathbf{x}$ are continuous on $U \times R^n$. Again, the state $\mathbf{x} \in R^n$ and control $\mathbf{u} \in U \subset R^m$ satisfy the state equation

$$d\mathbf{x}/dt = \mathbf{f}(\mathbf{x}, \mathbf{u}), \tag{3.2}$$

where $\mathbf{f}(\mathbf{x}, \mathbf{u})$ and $\partial \mathbf{f}(\mathbf{x}, \mathbf{u})/\partial \mathbf{x}$ are continuous on $U \times R^n$.

In particular the bilinear equation (1.9) repeated for convenience is

$$d\mathbf{x}/dt = \mathbf{A}\mathbf{x} + \sum_{k=1}^{m} u_k \mathbf{B}_k \mathbf{x} + \mathbf{C}\mathbf{u}. \tag{3.3}$$

Again, $\mathbf{u}(t)$ is a piecewise continuous vector function, and U is a compact set.

Dynamic optimization problems traditionally have been solved by the calculus of variations [1, 2]. Despite its many shortcomings, this somewhat loosely collected theory can be applied to optimal control problems as shown by Hestenes [3], Merriam [4], and Leitmann [5]. During the past decade or so, however, unified optimal control theories, based on dynamic programming and on the maximum principle, have been developed by Bellman [6] and Pontryagin *et al.* [7], respectively. It is generally accepted that the maximum principle is more of a research tool than a technique for designing practical control systems and that the opposite is true for dynamic programming. Still, many preliminary designs are based on the maximum principle, and much research has been associated with dynamic programming. It is shown in this book that the maximum principle can be quite useful to obtain the form of optimal control, while programming techniques may be utilized to determine the actual optimal policy. Quasi-linearization [8], gradient techniques [5, 9, 10], linear programming [11, 12], and nonlinear programming [13] as well as dynamic programming are frequently effective for computation of optimal control policies. Optimal control theory is related to mathematical programming techniques by Cannon *et al.* [14]. Similarly, Pontryagin *et al.* [7] and Kalman [15] show the relation of the maximum principle, and Dreyfus [16] shows the relation of dynamic programming to the calculus of variations.

Frequently, it is desired to find the control that steers the system (1.1) from an initial state $\mathbf{x}(t_0) = \mathbf{x}_0$ to some terminal state $\mathbf{x}(t_f) = \mathbf{x}_f$ so as to minimize (3.1) with an admissible control. Any control which does this is said to be optimal. While mathematical problems of existence and uniqueness of optimal solutions may be extremely complex some theorems are available, and physical arguments sometimes may be quite useful [7, 15, 17–19]. Lee and Markus [17] provide an existence theorem for a class of systems linear in control which is defined by

$$d\mathbf{x}/dt = \mathbf{a}(\mathbf{x}) + \mathbf{B}(\mathbf{x})\mathbf{u}, \tag{3.4}$$

where $\mathbf{a}(\mathbf{x})$ and $\mathbf{B}(\mathbf{x})$ are continuous functions with continuous derivatives. Obviously, this system is a generalization of the bilinear system. The performance index (3.1) considered here also is linear in control with

$$f_0(\mathbf{x}, \mathbf{u}) = a_0(\mathbf{x}) + \langle \mathbf{b}_0(\mathbf{x}), \mathbf{u} \rangle. \tag{3.5}$$

Theorem Let the system be defined by (3.4) with (a) an initial state $\mathbf{x}_0 \in R^n$ and a continuously moving compact target set $\theta_f \subset R^n$ on the finite interval $[t_0, t_f]$; (b) a performance index given by (3.1) and (3.5); and (c) controllability such that a nonempty set of admissible controls causes $\mathbf{x}(t_0) = \mathbf{x}_0$ to connect with $\mathbf{x}(t_f) \in \theta_f$ by uniformly bounded trajectories. Then there exists an admissible optimal control which steers (3.4) from $\mathbf{x}_0$ to θ_f so as to minimize (3.1) and (3.5).

Several optimal control problems for systems linear in control are analyzed by Athans and Falb [20], Hermes and Haynes [21], Kelley [22], and Hermes [23]. As an indication of the significance of this class of systems, Balakrishnan [24] shows that under somewhat general conditions the dynamics of controllable nonlinear processes can be specified by Eq. (3.4).

Application of the maximum principle shows immediately that the form of optimal control is a bang–bang process. If U is defined by $|u_k| \leq 1$, and $s_k(t)$ is the switching function, the control has the form of

$$u_k = \operatorname{sgn} s_k(t), \qquad k = 1, \ldots, m, \tag{3.6}$$

where $\operatorname{sgn}(s_k) = s_k/|s_k|$ for $s_k \neq 0$. If $s_k(t) = 0$ for some finite time, the controls are said to be singular and these must be considered. Optimal bang–bang controlled systems linear in control and particularly bilinear systems are analyzed in Section 3.3. There a gradient algorithm is derived to compute optimal switching times. Now, however, optimal bilinear regulation is studied. The discussion follows that reported by Mohler and Rink [25, 26].

3.2 OPTIMAL BILINEAR REGULATION

The linear regulator problem is analyzed at length in the literature [17, 20, 26]. While regulation of a linear system has been conventionally accomplished by linear feedback of the error signals which have been multiplied by constant gains, it is shown here that performance may be improved significantly by varying the feedback gains in some controlled manner. This results in bilinear regulation.

The regulation of bilinear systems (2.3) analyzed here, optimally transfers the process from some initial state to a small hypersphere about the origin in state space. An optimal synthesis of this regulator can follow a systematic procedure. By computing backward in time from terminal states, a family of extremal trajectories can be computed from the maximum principle and the necessary transversality condition of the costate vector. If there is an overlapping of certain extremal trajectories in state space, one covering of unique trajectories can be obtained by truncating trajectories while maintaining continuous cost surfaces; since there may be several such extremal coverings, however, the optimal policy is still unformulated. This procedure is demonstrated below by an example with overlapping extremal trajectories in a region.

To establish optimality, the results of Boltyanskii [19] may be utilized directly for the case in which the target is a single point, and in Appendix A this work is extended to accommodate problems with a smooth terminal hypersurface S. This extension is stated in the form of a theorem below.

Theorem If the exterior of S is partitioned into a finite number of connected cells by hypersurface elements whose union is designated by V, with $\partial \mathbf{u}(\mathbf{x})/\partial \mathbf{x}$ continuous on the interior of each cell, then an extremal control $\mathbf{u}(\mathbf{x})$ is optimal with respect to minimization of (3.1) if

(a) the extremal trajectory $\boldsymbol{\varphi}(\tau)$, transversing from $\mathbf{x}$ to S with control $\mathbf{u}[\boldsymbol{\varphi}(\tau)]$, intersects V with nonzero angle at a finite number of points and intersects S after an elapsed time $t(\mathbf{x})$;

(b) the cost to reach S from $\mathbf{x}$,

$$J(\mathbf{x}) = \int_0^{t(\mathbf{x})} f_0\{\boldsymbol{\varphi}(\tau), \mathbf{u}[\boldsymbol{\varphi}(\tau)]\}\, d\tau, \tag{3.7}$$

is a continuous function of $\mathbf{x}$ on all of R^n.

This theorem is used in the control system synthesis to be described next, where $\boldsymbol{\varphi}(\tau)$ is replaced by $\mathbf{x}(\tau)$ for convenience.

3.2.1 Quadratic-State Index

As an application of the foregoing theorem and optimal synthesis procedure consider the following second-order regulator with classical position feedback and controlled rate feedback:

$$\begin{aligned} \dot{x}_1 &= x_2 \\ \dot{x}_2 &= -x_1 - \alpha(1 + u)x_2, \end{aligned} \tag{3.8}$$

where x_1 is a position error and $|u| \leq 1$. The optimal control is to transfer an initial state to a small circle of radius ρ about the origin in such a manner as to minimize

$$J(\mathbf{x}) = \int_{-t(\mathbf{x})}^{0} [x_1^2(\tau) + x_2^2(\tau)]\, d\tau. \tag{3.9}$$

A family of extremal trajectories can be generated by running both the state and costate systems backward in time from "initial" points $\mathbf{x}(0)$ on S, provided that initial conditions can be established for $\mathbf{p}(0)$. Here $\mathbf{p}(t)$ is the adjoint state used in conjunction with the maximum principle or generalized Lagrange multipliers with the calculus of variations [17, 20, 26].

If $\{\mathbf{x}(\tau) \mid \tau \in [-t(\mathbf{x}), 0]\}$ is an extremal trajectory from $\mathbf{x}$ to $\mathbf{x}(0) \in S$, the maximum principle requires that there exist a costate vector function $\mathbf{p}(\tau)$ which is defined by the adjoint system [26] of (3.8) with $\mathbf{p}(0)$ normal to S at $\mathbf{x}(0)$. In the present case, $\mathbf{p}(0)$ must be parallel to the vector $\mathbf{x}(0)$ since S is a circle. Thus, set $\mathbf{p}(0) = \lambda\mathbf{x}(0)$, where λ is a negative number which remains to be specified.

The value of $u(\tau)$ along an extremal, as deduced from the maximum principle, is

$$u(\tau) = -\operatorname{sgn}[p_2(\tau)x_2(\tau)]. \tag{3.10}$$

Suppose there exists a singular arc, that is, a segment of extremal along which $p_2(\tau)\cdot x_2(\tau) = 0$ for $\tau \in [\tau_a, \tau_b]$. This implies that either $x_2(\tau) = \dot{x}_2(\tau) = 0$ or $\dot{p}_2(\tau) = p_2(\tau) = 0$ for $\tau \in [\tau_a, \tau_b]$. The former case is impossible since Eqs. (3.8) would imply that $x_1(\tau) = 0$ as well, and the origin in state space is not exterior to S. In the latter case, however, the adjoint system yields $\dot{p}_1 = 2x_1$ and $p_1 = -2x_2$. Therefore

$$\begin{aligned} \dot{x}_2(\tau) &= x_1(\tau) = -x_1(\tau) - \alpha(1 + u(\tau))x_2(\tau) \\ x_1(\tau) &= -\tfrac{1}{2}\alpha(1 + u(\tau))x_2(\tau) \end{aligned} \tag{3.11}$$

along the singular arc. Also, substitution of $p_2(\tau) = 0$ and $p_1(\tau) = -2x_2(\tau)$ into the Hamiltonian $H = 0$, where

$$H(\mathbf{x}, \mathbf{p}, \mathbf{u}) = \langle \mathbf{f}(\mathbf{x}, \mathbf{u}), \mathbf{p} \rangle + f_0(\mathbf{x}, \mathbf{u})p_0$$

with $p_0 = -1$ for convenience, yields the further condition

$$x_1^2(\tau) = x_2^2(\tau).$$

If $x_1(\tau) = x_2(\tau)$, (3.11) implies that $u(\tau) < -1$, which is not in U. Thus the only singular arcs are on the line $x_1(\tau) = -x_2(\tau)$, and the control value along such an arc is

$$u(\tau) = (2/\alpha) - 1,$$

which is in U for $\alpha \geq 1$.

The extremals which are generated by solving the reversed time state and costate equations for initial states $\mathbf{x}(0)$ on S, with u determined by (3.10), do not contain any of the singular arcs above. Also, it is shown below that these extremals satisfy sufficient conditions for optimality, and hence they are optimal. It follows that an extremal containing a singular arc has, at best, an equal cost functional.

It is necessary to evaluate λ in order to determine the initial costate vector $\mathbf{p}(0) = \lambda\mathbf{x}(0)$. Since $p_2(0) = \lambda x_2(0)$, (3.10) requires that $u(0) = +1$. Then $H = 0$ if and only if

$$\lambda = [x_1^2(0) - x_2^2(0)]/2\alpha x_2^2(0) \tag{3.12}$$

which exists almost everywhere on S.

The families of extremals for $\alpha = 1$ and $\alpha = \frac{1}{2}$, obtained with an analog computer, are shown in Figs. 3.1 and 3.2. In each case the extremals cover the entire plane and do not intersect one another, and the corresponding cost functional is continuous.

In each figure, the plane is divided by "switching curves" into regions on which u is constant. These switching curves constitute the set V, and it is evident that the extremals cross V with nonzero angles. In each case the $x_2 = 0$ axis is a switching curve because of (3.10). For

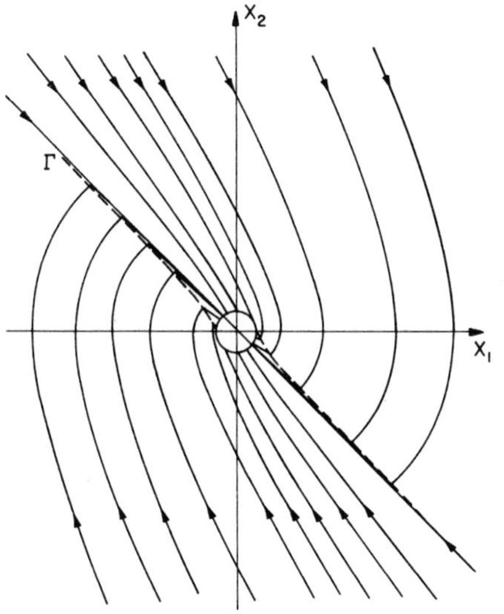

Fig. 3.1. Optimal bilinear regulation, $\alpha = 1$.

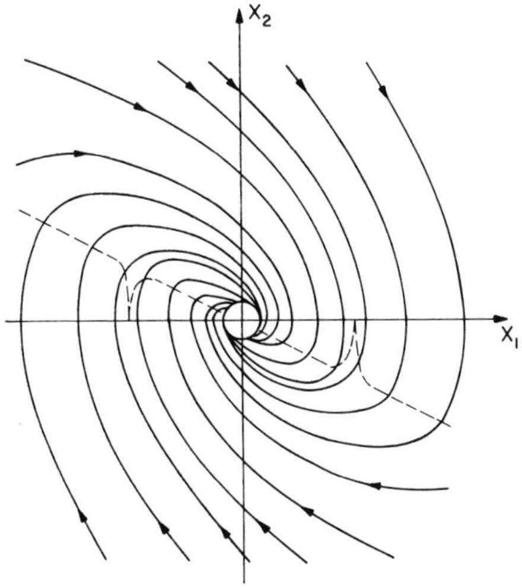

Fig. 3.2. Optimal bilinear regulation, $\alpha = \frac{1}{2}$.

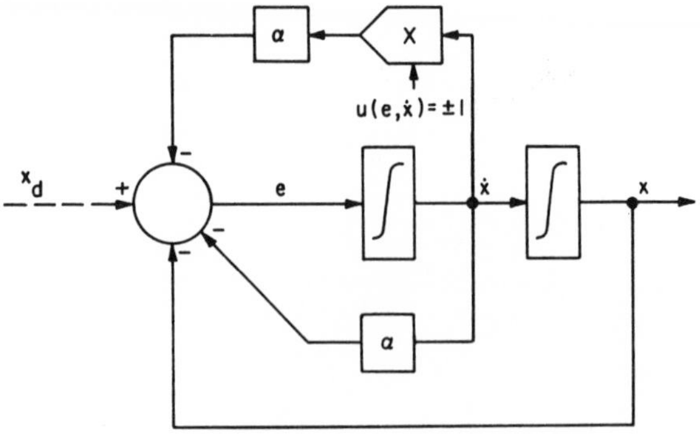

Fig. 3.3. Bilinear regulator synthesis, $n = 2$.

the case $\alpha = \frac{1}{2}$, the eigenvalues of the system with $u = +1$ are complex, and two switching curves are shown in Fig. 3.2 which have the looped characteristic of switching curves for linear systems with complex eigenvalues. The control system state diagram is shown in Fig. 3.3.

It is of interest to examine the effect of changing the radius ρ on the extremal synthesis. Suppose that $\mathbf{x}(\tau)$ is any extremal terminating on a circle S_1 with radius ρ_1, and that $\mathbf{p}(\tau)$ is the corresponding costate trajectory, with $\mathbf{p}(0) = \lambda\mathbf{x}(0)$. The control function $u(\tau)$ along the trajectory, as given by (3.10), is $u(\tau) = -\operatorname{sgn}[x_2(\tau)p_2(\tau)]$.

Note that the state–costate system is linear and homogeneous, and that expression (3.12) for λ is dimensionless. Moreover, the switching function $p_2(\tau)x_2(\tau)$ is homogeneous. Therefore $\mathbf{x}(\tau)$ and $\mathbf{p}(\tau)$ can be magnitude scaled by a constant factor ρ_2/ρ_1, and the result is an extremal which terminates on the circle S_2 with radius ρ_2.

Now consider the synthesis of Fig. 3.1. The extremals corresponding to a circle of radius $k\rho$ are obtained from those shown by performing a uniform contraction (dilation) with scale factor k on the entire phase plane. In particular, the switching curve Γ of Fig. 3.1 is mapped into a new switching curve $\Gamma_k = \{\mathbf{x} \mid \mathbf{x}/k \in \Gamma\}$.

3.2.2 Suboptimal Servomechanism

Suppose that the target circle undergoes an infinite contraction, that is, $k \to 0$. In the limit, S becomes the origin in the phase plane. As was mentioned previously, the optimization problem is not well posed when

S is just the origin since this point is not reachable in finite time. However, the switching curves Γ_k obtained by the contraction process approach a well-defined limit. As $k \to 0$ the switching curve $\Gamma_k = \{\mathbf{x} \mid \mathbf{x}/k \in \Gamma\}$ becomes just the radial line with slope $-\lambda_1$. Thus the limiting extremal synthesis is shown by Fig. 3.4. The plane is divided into sectors R^+ and R^-, with $u(\mathbf{x}) = +1$ on R^+ and $u(\mathbf{x}) = -1$ on R^-. Again, Fig. 3.3 shows the system state diagram.

The result can be given a very simple interpretation. The eigenvector $\mathbf{w}^2$ corresponding to the smaller eigenvalue λ_2 lies in the dashed line shown in Fig. 3.3. Without the bilinear mode of control (i.e., for the equation $\ddot{x} + 2\alpha\dot{x} + x = 0$), all trajectories would approach the origin asymptotic to the direction of the "slow" eigenvector $\mathbf{w}^2$. This would result in relatively poor performance. Moreover, as α becomes large the performance deteriorates, since $\lambda_2 \to 0$. With the bilinear mode of control, however, the direction of $\mathbf{w}^2$ lies entirely in the sectors R^-, and the effect of the "slow mode" is nullified. All extremals approach the origin along the "fast" eigenvector $\mathbf{w}^1$. As the damping α becomes large, the eigenvalue λ_1 becomes large and the performance improves monotonically. Unfortunately, it can be shown that the limiting synthesis of Fig. 3.4 is not quite optimal.

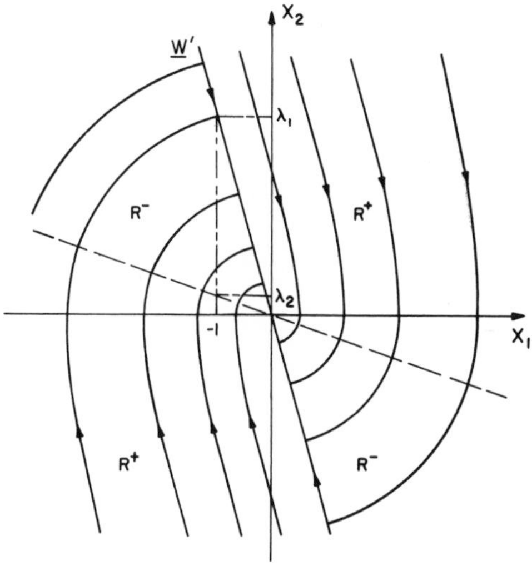

Fig. 3.4. Limiting synthesis, $\alpha = 2$.

It is of interest to compare the performance of the second-order bilinear system with that of a linear system, when the input is other than a step. Hence a sinusoidal input with variable frequency is utilized for the bilinear system

$$\ddot{x} + \alpha[1 + u(x - x_{\mathrm{d}}, \dot{x})]\dot{x} + x = x_{\mathrm{d}}, \tag{3.13}$$

where x is the system response for a demanded input x_{d}. The simple switching law shown in Figs. 3.3 and 3.4 is incorporated for $u(x - x_{\mathrm{d}}, \dot{x})$. This law was obtained as the limit of syntheses for step function inputs, and was shown to be suboptimal even for this class of inputs. Therefore no claim is made as regards optimality for the results which are presented.

The law of Fig. 3.4, however, has two pleasant properties. In the first place, it is very easy to synthesize in practice. Also, the nonlinearity is strictly independent of radius in the phase plane, which means that solutions of the resulting system have the important scalar multiplication property. That is, if the forced system (3.13) has the response $x(t)$ for an input $x_{\mathrm{d}}(t)$, it has response $kx(t)$ for the input $kx_{\mathrm{d}}(t)$. Observe that because the switching lines in Fig. 3.4 are radial,

$$u(x - x_{\mathrm{d}}, \dot{x}) = u(kx - kx_{\mathrm{d}}, k\dot{x}),$$

and therefore

$$k\ddot{x} + \alpha[1 + u(kx - kx_{\mathrm{d}}, k\dot{x})]k\dot{x} + kx = kx_{\mathrm{d}}$$

is satisfied.

The solid curves of Fig. 3.5 represent normalized mean-squared

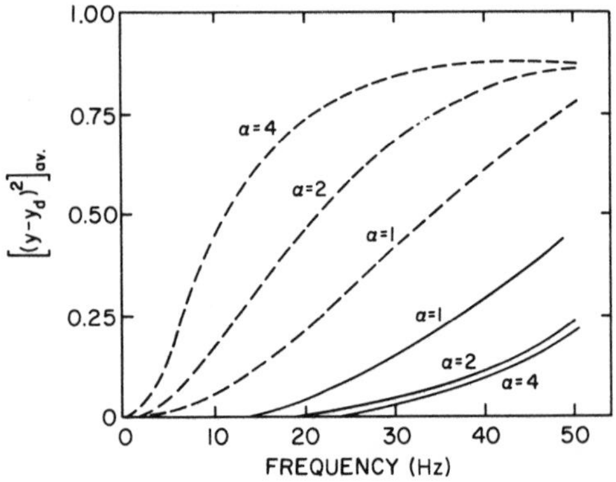

Fig. 3.5. Performance of linear versus bilinear servosystems for sinusoidal input.

tracking error† as a function of frequency for the bilinear system with $\alpha = 1, 2,$ and 4. Because of the second property discussed previously, these are universal curves which do not depend on the amplitude of $x_d(t)$. The dashed curves represent mean-squared error for the linear system obtained by setting $u = 1$ in Eq. (3.13), with $\alpha = 1, 2,$ and 4. The performance of the bilinear system is clearly superior, and the trend for increasing α is reversed from that of the linear system.

3.2.3 Time-Optimal Regulation

The synthesis of a time-optimal control law to transfer the state of (3.8) to some circle about the origin in state space is now analyzed.

The Hamiltonian for the minimum-time case is

$$H(\mathbf{x}, \mathbf{p}, u) = p_0 + p_1 x_2 - p_2[x_1 + \alpha(1 + u)x_2],$$

and the costate or adjoint system is

$$\begin{aligned} \dot{p}_1 &= p_2 \\ \dot{p}_2 &= -p_1 + \alpha(1 + u)p_2. \end{aligned} \tag{3.14}$$

Along an extremal trajectory, the Hamiltonian is maximized when the control value is

$$u = -\operatorname{sgn}(p_2 x_2). \tag{3.15}$$

It is easily verified that no singular arcs exist for this problem.

Now, extremals are generated by computing backward in time from initial points $(x_1(0), x_2(0))$ on a target circle about the origin in the phase plane. Since the normal to a circle is its radius vector, the transversality condition of the maximum principle is satisfied if we set $p_1(0) = -x_1(0)$, $p_2(0) = -x_2(0)$. Then the value of the Hamiltonian is zero if the constant p_0 is equal to $-2\alpha x_2^2(0)$, which is negative as required by the maximum principle.

The radius of the target circle can, without loss of generality, be taken as unity, since systems (3.8) and (3.14) and relation (3.15) are homogeneous in $\mathbf{x}$ and $\mathbf{p}$. Again, the family of extremals for a target circle of arbitrary radius k can be obtained from those for the unit circle by performing a uniform contraction or dilation with scale factor k in the phase plane.

† Here mean-squared tracking error refers to a time-integrated average of the square of the difference between the desired response and the actual response.

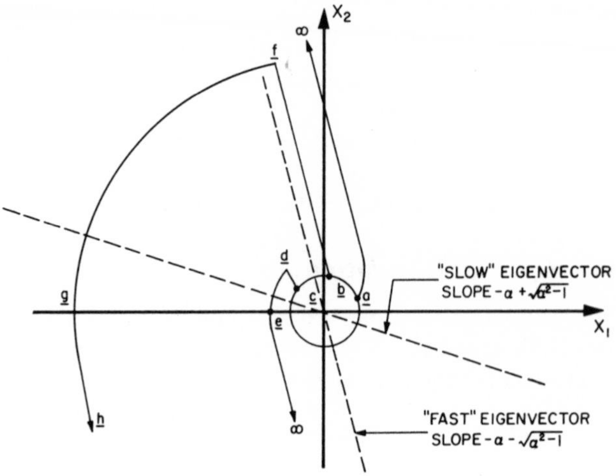

Fig. 3.6. Qualitative phase description for time-optimal regulator.

Qualitative aspects of the extremals for the case $\alpha \geq 1$ are easily deduced. Three extremals are indicated on the phase plane shown in Fig. 3.6, while the corresponding trajectories in the costate plane are shown in Fig. 3.7. The eigenvectors for systems (3.8) and (3.14) with $u = 1$ are also shown. For the initial state **a**, the corresponding initial costate is **a′**, and the costate trajectory tends toward the origin without crossing the p_1 axis as time approaches infinity. Thus no switching occurs, and the extremal trajectory from **a** goes to infinity as shown.

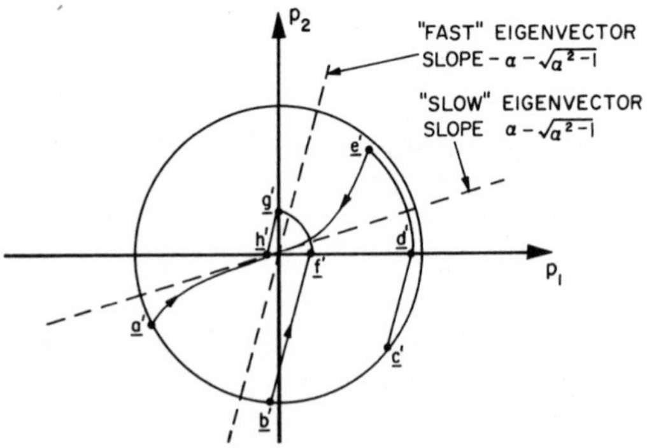

Fig. 3.7. Adjoint trajectories corresponding to Fig. 3.6.

For the initial state **c**, the costate trajectory from **c**′ crosses the p_1 axis at **d**′, causing u to switch to -1. The subsequent motions of state and costate are along the circular arcs from **d** to **e** and **d**′ to **e**′, respectively, where the angles subtended by these arcs are equal. At **e**, $x_2 = 0$, and the switching law (3.15) indicates that u switches back to $+1$. Thereafter, p_2 never crosses zero again, so no further switchings occur, and the extremal trajectory approaches infinity as shown on Fig. 3.6.

Along the extremal from the initial state **b**, two analogous switchings occur at **f** and **g**, as shown. The angle subtended by the circular arc from **f** to **g**, however, is greater than that which the steeper eigenvector makes with the negative x_1 axis. Therefore, in rotating from **f**′ to **g**′ the costate vector passes through the steeper eigenvector as well. Consequently, a third switching will occur when the costate vector reaches **h**′, and the subsequent circular arc from the distant switching point **h** will cross the extremal from **c**.

Thus certain extremals cross one another, and it is necessary to perform a truncation in order to obtain a simple covering of the phase plane and a unique control law. It is desired, however, that the resulting control law satisfy the sufficient conditions for optimality given in the previous section. In particular, according to condition (c), the truncation should be done in such a way that the cost functional is everywhere continuous.

For two-dimensional problems, it is not difficult to compute and plot isocost curves for each of the intersecting subfamilies in the region where they cross. The point of intersection of isocost curves for each given cost then fixes a point on the decision boundary, on either side of which the subfamily with higher cost is truncated. The result is a continuous, single-valued cost functional.

The result of this procedure, for the present problem with $\alpha = 2$, is shown in Fig. 3.8. Only those extremals that switch and tend to cross each other in this region of the fourth quadrant are shown, and they are truncated along the upper right boundary of the $u = -1$ cell so that the cost functional (time, in this case) is continuous. The dotted lines are isochrones for the indicated time values. The lower left boundary of the $u = -1$ cell and the dashed curve shown inside this cell are the loci of the third switching mentioned above.

The optimal switching law for this problem with $\alpha = 2$ is shown in Fig. 3.9, where the unit circle cannot be shown because of the scale. The

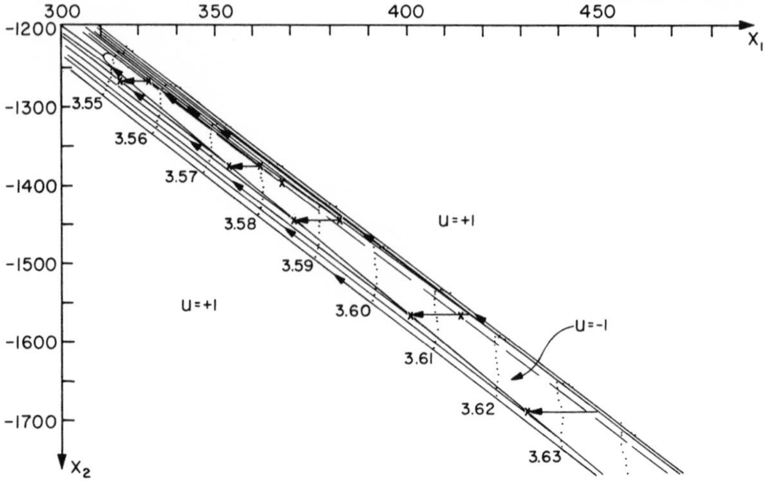

Fig. 3.8. Time-optimal phase trajectories showing truncation.

existence of the large cells with $u = -1$ can be given a simple interpretation. The "slow" eigenvector of the system (3.8) with $u = +1$ lies in these cells, and in order to conserve time the phase should move with zero damping ($u = -1$) to the vicinity of the fast eigenvector before moving toward the origin with maximum damping.

These considerations can be extended to higher order regulator systems. In classical feedback design, it is well known that rate feedback can stabilize many systems and improve their transient response. On the other hand, an excessive amount of rate feedback causes certain poles (eigenvalues) to approach zero, resulting in a degradation of transient response. Again, this limitation can be circumvented with a bilinear mode of control, since the trajectory can be made to approach its equilibrium state along fast eigenvectors, avoiding those slow ones which correspond to the small eigenvalues. The optimum control law can be determined by the method described here, although the difficulty involved in fitting cost surfaces together in a continuous fashion will certainly increase rapidly with the order of the system.

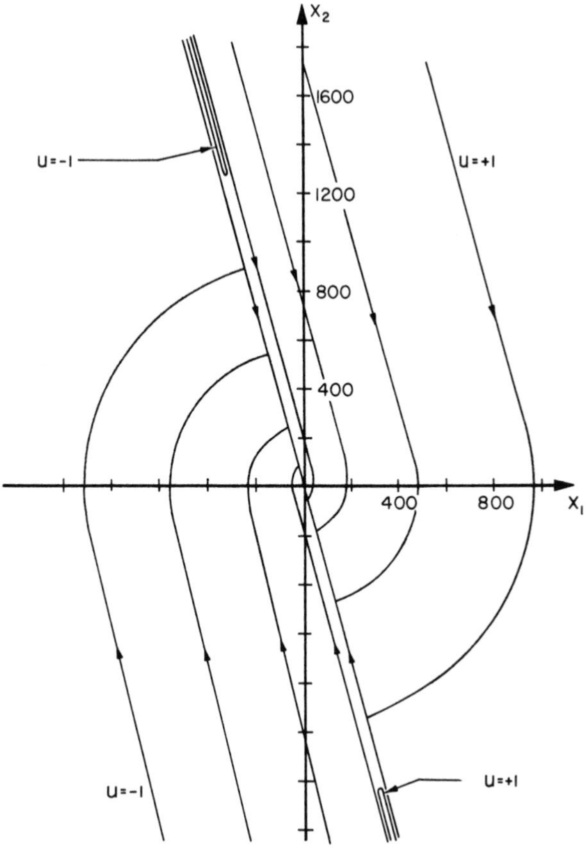

Fig. 3.9. Time-optimal phase trajectories, $\alpha = 2$.

3.3 BANG–BANG PROCESSES

The optimal control of systems linear in control (SLIC) (3.4) with respect to linear-in-control performance (3.5) is studied in this section. As indicated in Section 3.1, it is obvious from the maximum principle that the form of optimal control, if it exists, is described by (3.6).

Singular solutions are neglected in this section so that the control can be synthesized by a bang–bang process (switching control). Also, it is assumed that the optimal control is to steer the system from some initial state $\mathbf{x}_0$ to some smooth terminal q-fold θ_f described by

$$\theta_f = \{\mathbf{x} \mid \boldsymbol{\psi}(\mathbf{x}) = \mathbf{0}\}, \tag{3.16}$$

where $\boldsymbol{\psi}(\mathbf{x}) \in R^q$. For this problem it is readily seen from the integral solution of (3.4) and the triangle inequality applied to the appropriate norm of both sides of the integral equation that the connecting $\mathbf{x}(t)$ is uniformly bounded [27]. Hence, controllability from the initial state to the terminal manifold yields existence of an optimal control. Simple problems sometimes may be solved by means of the maximum principle and its associated two-point boundary value problem. For the addended SLIC with

$$d\tilde{\mathbf{x}}/dt = \tilde{\mathbf{a}}(\tilde{\mathbf{x}}) + \tilde{\mathbf{B}}(\tilde{\mathbf{x}})\mathbf{u}, \tag{3.17}$$

where

$$\tilde{\mathbf{x}} = \begin{bmatrix} x_a \\ \mathbf{x} \end{bmatrix}, \qquad \tilde{\mathbf{x}}(t_0) = \begin{bmatrix} 0 \\ \mathbf{x}_0 \end{bmatrix}, \qquad \tilde{\mathbf{B}}(\tilde{\mathbf{x}}) = \begin{bmatrix} \mathbf{b}_0^{\mathrm{T}}(\mathbf{x}) \\ \mathbf{B}(\mathbf{x}) \end{bmatrix},$$

$$\tilde{\mathbf{a}}(\tilde{\mathbf{x}}) = \begin{bmatrix} a_0(\mathbf{x}) \\ \mathbf{a}(\mathbf{x}) \end{bmatrix}, \qquad \text{and} \qquad x_a(t_f) = J.$$

For the maximum principle

$$H(\tilde{\mathbf{x}}, \tilde{\mathbf{p}}, \mathbf{u}) = \langle \tilde{\mathbf{p}}, \tilde{\mathbf{a}}(\tilde{\mathbf{x}}) + \tilde{\mathbf{B}}(\tilde{\mathbf{x}})\mathbf{u} \rangle \tag{3.18}$$

is maximized by (3.6) or

$$\mathbf{u} = \mathrm{SGN}(\mathbf{s}(t)),$$

where

$$\mathbf{s}(t) = \mathbf{B}^{\mathrm{T}}(\mathbf{x})\mathbf{p}, \tag{3.19}$$

and SGN operates on each vector component.

Here, the costate is described by

$$d\tilde{\mathbf{p}}/dt = -\partial H/\partial \tilde{\mathbf{x}}, \tag{3.20}$$

where, to specify a two-point boundary value problem (TPBVP), $\tilde{\mathbf{p}}(t_f)$ is orthogonal to the tangent plane of the smooth terminal manifold at $\tilde{\mathbf{x}}(t_f)$.

Usually, however, it is not possible to obtain an analytical solution to this TPBVP, and numerical techniques are utilized. A common procedure is initially to guess a solution which satisfies some boundary conditions, and then to solve the TPBVP recursively so that the remaining conditions are solved more and more closely in each iteration. Next, an iterative computational method, called the switching-time-variation method (STVM), is developed for a successive approximation of a solution to the TPBVP.

3.3.1 The Switching-Time-Variation Method (STVM)

The concept of switching-time-variation is introduced in order to compute the gradient of the cost functional with respect to switching times. Here the work of Mohler and Moon [27] is followed.

The STVM is a computational technique to obtain optimum bang–bang controls for the problem discussed above by means of a modified gradient scheme. The basic idea of the method is to generate a sequence of switching functions and to compute the gradient of the cost with respect to the switching times. With this gradient, the switching times are corrected in each iteration so that the control sequence converges to an optimal nonsingular control.

For convenience, it is assumed that the controls are scalar functions, but the arguments can be generalized easily for vector controls. Then (3.17) can be replaced by

$$d\tilde{\mathbf{x}}/dt = \tilde{\mathbf{a}}(\tilde{\mathbf{x}}) + \tilde{\mathbf{b}}(\tilde{\mathbf{x}})u, \tag{3.21}$$

where

$$\tilde{\mathbf{b}}(\tilde{\mathbf{x}}) = \begin{bmatrix} b_0(\mathbf{x}) \\ \mathbf{b}(\mathbf{x}) \end{bmatrix}.$$

Also, a free terminal state (i.e., $\theta_{\mathrm{f}} = R^n$) and a fixed terminal time are assumed to simplify the gradient computations. This does not mean, however, that the method cannot be utilized for problems with a constrained terminal set θ_{f}. For these problems, the target set can be considered by means of adjoining a penalty term to the cost functional so that the problem with modified cost converges to the original problem [28]. Free terminal time problems may be solved by means of a sequence of fixed-time problems.

As a basis for the method, define the switching vectors $\boldsymbol{\tau} \in R^N$ and $\boldsymbol{\tau}_i \in R^N$, and the switching set $T \subset R^N$ so that

$$\boldsymbol{\tau}^N = (\tau_1{}^N, \ldots, \tau_N{}^N)^{\mathrm{T}} \tag{3.22}$$

$$\boldsymbol{\tau}_i{}^N = (\tau_{i1}^N \ \ldots, \tau_{iN}^N)^{\mathrm{T}}, \tag{3.23}$$

and

$$T^N = \{\boldsymbol{\tau}^N: t_0 \leq \tau_1{}^N \leq \cdots \leq \tau_N{}^N \leq t_{\mathrm{f}}\}. \tag{3.24}$$

The elements $\tau_j{}^N$ and τ_{ij}^N $(j = 1, \ldots, N)$ of $\boldsymbol{\tau}^N$ and $\boldsymbol{\tau}_i{}^N$ are called switching times. Also, let

$$v(\boldsymbol{\tau}^N) = \sum_{j=0}^{N} (-1)^j[\delta_{-1}(t - \tau_j{}^N) - \delta_{-1}(t - \tau_{j+1}^N)]$$

and

$$V^N = \{v(\boldsymbol{\tau}^N): \boldsymbol{\tau}^N \in T^N, N = \text{nonnegative integers}\},$$

where $\delta_{-1}(t)$ is the unit step function. It follows that $V^N \subset U$, and a bang–bang control $u(t) \in V$ such that

$$u(t) = v(\boldsymbol{\tau}^N) \tag{3.25}$$

for some N and some $\boldsymbol{\tau}^N \in T^N$.

Substitution of (3.25) into (3.21) yields

$$d\tilde{\mathbf{x}}/dt = \tilde{\mathbf{a}}(\tilde{\mathbf{x}}) + (-1)^j\tilde{\mathbf{b}}(\tilde{\mathbf{x}}) \tag{3.26}$$

for $t \in [\tau_j{}^N, \tau_{j+1}^N)$, $j = 0, \ldots, N$, where $\tau_{N+1}^N = t_{\mathrm{f}}$. Obviously, (3.26) has a continuous and unique solution $\tilde{\mathbf{x}}(t)$ on $[\tau_j{}^N, \tau_{j+1}^N)$ from which it can be shown that $\tilde{\mathbf{x}}(t_{\mathrm{f}})$, and therefore $x_a(t_{\mathrm{f}}) = J$, is continuous in $\boldsymbol{\tau}^N$ [30].

Then define a continuous function I by

$$I(\boldsymbol{\tau}^N) = J\big(\mathbf{x}_0, v(\boldsymbol{\tau}^N), t_0\big) = x_a(t_{\mathrm{f}}). \tag{3.27}$$

Since T^N is compact in R^N, $I(T^N)$ is compact, and hence there exists a switching vector $\boldsymbol{\tau}^{N*}$ in T^N and an element I^{N*} such that

$$I^{N*} = \min_{\boldsymbol{\tau}^N \in T^N} I(\boldsymbol{\tau}^N) = I(\boldsymbol{\tau}^{N*}). \tag{3.28}$$

Also, since $T^N \subset T^{N+1}$,

$$I^{N*} = \min_{\boldsymbol{\tau}^N \in T^N} I(\boldsymbol{\tau}^N) \geq \min_{\boldsymbol{\tau}^{N+1} \in T^{N+1}} I(\tau^{N+1}) = I^{N+1*}.$$

Consequently, $\{I^{N*}\}$ is a nonincreasing sequence, and obviously $I^{N*} \geq$

I° for all N, where $I^\circ = \min_{u \in V} J(x_0, u, t_0)$. (Here, existence of an optimal $u \in v$ control is assured by the theorem above.) Since the nonincreasing sequence $\{I^{N^*}\}$ is bounded below by I°, there exists a number $I^* \geq I^\circ$ such that

$$\lim_{N \to \infty} \{I^{N^*}\} = I^* = \inf_N \{I^{N^*}\}. \tag{3.29}$$

It is readily seen that a nonsingular bang–bang control has a finite number of switchings [29]. Then there exists a nonnegative integer N_o and a switching vector $\boldsymbol{\tau}^{N_o*} \in T^{N_o}$ such that $u^\circ(t) = v(\tau^{N_o*})$ and

$$I^{N_o*} \leq I^{N*} \qquad \text{for all} \qquad N \geq N_o. \tag{3.30}$$

Next it is shown how to compute the switching vector $\boldsymbol{\tau}^{N*}$ which corresponds to the minimum element I^{N*} for each given N. Then a method is derived in Appendix B to compute the optimum number of switchings N_o so that (3.30) is satisfied, and finally it is shown that the limit of the sequence $\{I^{N^*}\}$ is equal to the optimal cost I°.

As shown in Appendix C, the gradient of I with respect to the switching vector $\boldsymbol{\tau}^N$ may be computed from

$$\operatorname{grad} I(\boldsymbol{\tau}_i^N) = (\partial I/\partial \boldsymbol{\tau}^N)(\boldsymbol{\tau}_i^N) = -\boldsymbol{\phi}_i^N, \tag{3.31}$$

where

$$\boldsymbol{\phi}_i^N = (\phi_{i1}^N, -\phi_{i2}^N, \ldots, (-1)^{N-1}\phi_{iN}^N)^{\mathrm{T}},$$

$$\phi_{ij}^N = \phi_i(\tau_{ij}^N),$$

$$\phi_i(t) = 2\langle \tilde{\boldsymbol{\lambda}}_i(t), \tilde{\mathbf{b}}(\tilde{\mathbf{x}}_i(t))\rangle,$$

and $\tilde{\boldsymbol{\lambda}}_i(t)$ is the solution to

$$\frac{d\tilde{\boldsymbol{\lambda}}_i}{dt} = -\frac{\partial}{\partial \tilde{\mathbf{x}}} [\tilde{\mathbf{a}}(\tilde{\mathbf{x}}_i) + v_i \tilde{\mathbf{b}}(\tilde{\mathbf{x}}_i)]^{\mathrm{T}} \tilde{\boldsymbol{\lambda}}_i,$$

with

$$\tilde{\boldsymbol{\lambda}}_i(t_f) = \left.\frac{\partial I}{\partial \tilde{\mathbf{x}}}\right|_{\tilde{\mathbf{x}}_i(t_f)}.$$

Now the gradient vector may be utilized to generate a nonincreasing sequence of cost functionals and corresponding switching vectors. In this manner let an improved switching vector be given by

$$\boldsymbol{\tau}_{i+1}^N = \boldsymbol{\tau}_i^N + \mathbf{K}_i \boldsymbol{\phi}_i^N, \tag{3.32}$$

where $\mathbf{K}_i$ is an $N \times N$ diagonal matrix with sufficiently small nonnegative elements such that $\boldsymbol{\tau}_{i+1}^N$ belongs to the switching set T^N. Then $I_{i+1}^N = I(\boldsymbol{\tau}_{i+1}^N)$ can be approximated by

$$\begin{aligned} I_{i+1}^N &= I_i^N + \langle \text{grad}\, I(\boldsymbol{\tau}_i^N), \mathbf{K}_i \boldsymbol{\phi}_i^N \rangle \\ &= I_i^N - \langle \boldsymbol{\phi}_i^N, \mathbf{K}_i \boldsymbol{\phi}_i^N \rangle \le I_i^N. \end{aligned} \tag{3.33}$$

Of course, the sequence $\{I_i^N\}$ generated by (3.33) is nonincreasing. Since it is bounded below by a minimum value I^{N*}, the sequence is convergent with some limit $\hat{I}^N \ge I^{N*}$. Then, since I is continuous on T^N, and $\boldsymbol{\tau}_i^N$ for each i belongs to T^N, there exists $\hat{\boldsymbol{\tau}}^N$ in T^N such that [3]

$$I(\hat{\boldsymbol{\tau}}^N) = \hat{I}^N. \tag{3.34}$$

For this switching set,

$$\mathbf{K}\boldsymbol{\phi}(\hat{\boldsymbol{\tau}}^N) = \mathbf{0}. \tag{3.35}$$

Otherwise, $\delta\hat{\boldsymbol{\tau}}^N = \mathbf{K}\boldsymbol{\phi}(\hat{\boldsymbol{\tau}}^N)$ could be nonzero such that $I(\hat{\boldsymbol{\tau}}^N + \delta\hat{\boldsymbol{\tau}}^N) < I(\hat{\boldsymbol{\tau}}^N)$, which is a contradiction.

While only a local minimum cost is attained for a given number of switchings N, it is shown in Appendix B how to compute the optimal number of switchings.

To show convergence of the STVM, let

$$w(\boldsymbol{\tau}_i^N) = I(\boldsymbol{\tau}_i^N) - I(\hat{\boldsymbol{\tau}}^N), \tag{3.36}$$

where $I(\boldsymbol{\tau})$ and $\hat{\boldsymbol{\tau}}^N$ are given by Eqs. (3.27) and (3.34), respectively. Then it can be seen that

$$w(\boldsymbol{\tau}) \ge 0 \qquad \text{for all} \quad \boldsymbol{\tau} \in \{\boldsymbol{\tau}_i^N\} \subset T^N \tag{3.37}$$

and that

$$\begin{aligned} \Delta w(\boldsymbol{\tau}_i^N) &= w(\boldsymbol{\tau}_{i+1}^N) - w(\boldsymbol{\tau}_i^N) = I(\boldsymbol{\tau}_{i+1}^N) - I(\boldsymbol{\tau}_i^N) \\ &= I(\boldsymbol{\tau}_i^N) = -\langle \boldsymbol{\phi}(\boldsymbol{\tau}_i^N), \mathbf{K}\boldsymbol{\phi}(\boldsymbol{\tau}_i^N) \rangle \le 0, \end{aligned} \tag{3.38}$$

where $\phi_i(t) = 2\langle \tilde{\boldsymbol{\lambda}}_i(t), \tilde{\mathbf{B}}(\tilde{\mathbf{x}}_i(t)) \rangle$. Also, from (3.36) and (3.37), it follows that $K\boldsymbol{\phi}(\hat{\boldsymbol{\tau}}^N) = \mathbf{0}$, and $w(\hat{\boldsymbol{\tau}}^N) = \Delta w(\hat{\boldsymbol{\tau}}^N) = 0$. Then the point $\hat{\boldsymbol{\tau}}^N$, which satisfies the condition $\mathbf{K}\boldsymbol{\phi}(\hat{\boldsymbol{\tau}}^N) = \mathbf{0}$ and $w(\hat{\boldsymbol{\tau}}^N) = \Delta w(\hat{\boldsymbol{\tau}}^N) = 0$, also causes $\boldsymbol{\tau}_{i+1}^N - \boldsymbol{\tau}_i^N = 0$ in (3.32). Therefore, every initial point $\boldsymbol{\tau}^N \in \{\boldsymbol{\tau}_i^N\}$ must converge to $\hat{\boldsymbol{\tau}}^N$, and once the stationary point $\hat{\boldsymbol{\tau}}^N$ is reached the system cannot leave the point.

Since $\{\boldsymbol{\tau}_i^N\}$ converges to $\hat{\boldsymbol{\tau}}^N$, and $\tilde{\mathbf{x}}(t)$ and $\tilde{\boldsymbol{\lambda}}(t)$ continuously depend on the switching vector $\boldsymbol{\tau}$, $\tilde{\mathbf{x}}_i$ and $\tilde{\boldsymbol{\lambda}}_i$ must converge to $\tilde{\mathbf{x}}$ and $\tilde{\boldsymbol{\lambda}}$, respectively, where $\tilde{\mathbf{x}}$ and $\tilde{\boldsymbol{\lambda}}$ are the solutions of the state and the adjoint equations which correspond to $v(\hat{\boldsymbol{\tau}}^N)$.

Hence, $\phi_i^N(t) = 2\langle\tilde{\boldsymbol{\lambda}}_i, \tilde{\mathbf{b}}(\tilde{\mathbf{x}}_i)\rangle$ converges to $\hat{\phi}_N{}^N(t) = 2\langle\tilde{\boldsymbol{\lambda}}, \tilde{\mathbf{b}}(\tilde{\mathbf{x}})\rangle$. Also, the convergence of $\{\phi_i^N(t)\}$ implies the convergence of $\{\boldsymbol{\phi}_i^N\}$.

3.3.2 Relation to the Maximum Principle

Here it is shown that a solution obtained by the STVM satisfies the TPBVP derived from the maximum principle.

From Eqs. (3.18), (3.19), and (3.21), and from the maximum principle and the terminal transversality condition, it is seen that an optimal solution must satisfy the TPBVP which is given by (3.17) along with

$$\frac{d\tilde{\mathbf{p}}}{dt} = -\frac{\partial}{\partial \mathbf{x}}[(\tilde{\mathbf{a}}(\tilde{\mathbf{x}}) + \tilde{\mathbf{b}}(\tilde{\mathbf{x}})u)]^{\mathrm{T}}\tilde{\mathbf{p}}, \tag{3.39}$$

$$\tilde{\mathbf{p}}(t_f) = \left(p_0, \frac{\partial \psi^*}{\partial x_1}, \ldots, \frac{\partial \psi^*}{\partial x_n}\right)^{\mathrm{T}}, \tag{3.40}$$

and

$$u = \operatorname{sgn} s(t), \tag{3.41}$$

where $p_0 \leq 0$ is a constant, $s(t) = \langle\tilde{\mathbf{p}}(t), \tilde{\mathbf{b}}(\tilde{\mathbf{x}}(t))\rangle$, and

$$\psi^* = \sum_{i=1}^{q} \sigma_i \psi_i(x_i(t_f)). \tag{3.42}$$

It is easily verified that the state vector $\tilde{\mathbf{x}}(t)$ obtained by the STVM satisfies the state equation (3.21) along with its initial condition, and that the adjoint vector $\tilde{\boldsymbol{\lambda}}(t)$ satisfies the costate equation (3.39). Also, from Eq. (C.2) of Appendix C and the free terminal state problem with adjoined performance index of the form

$$I = \int_{t_0}^{t_f} [a_0(\mathbf{x}) + b_0(\mathbf{x})u]\, dt + \sum_{i=1}^{q} \rho_i \psi_i^2(\mathbf{x}(t_f)), \tag{3.43}$$

$$\tilde{\boldsymbol{\lambda}}(t_f) = \left.\frac{\partial I}{\partial \tilde{\mathbf{x}}}\right|_{t_f} = \left[\left.\frac{\partial I}{\partial x_0}\right|_{t_f}, \ldots, \left.\frac{\partial I}{\partial x_n}\right|_{t_f}\right]^{\mathrm{T}} \tag{3.44}$$

From (3.43) and (3.44), it follows that $\boldsymbol{\lambda}(t_f)$ is orthogonal to θ_f, and that $p_0\tilde{\boldsymbol{\lambda}}(t)$ satisfies (3.40).

Since the costate (3.39) is homogeneous and $\tilde{\boldsymbol{\lambda}}(t)$ satisfies the equation,

$p_0\tilde{\boldsymbol{\lambda}}(t)$ also satisfies (3.39). Therefore, $p_0\tilde{\boldsymbol{\lambda}}(t)$ satisfies both the costate and the terminal boundary condition.

From Eq. (B.5) of Appendix B, $Z_\tau^{\hat{R}} = Z_\phi^{\hat{R}}$. Then from the bang–bang control that results from the maximum principle and from (B.2) and (3.35) it follows that on $[t_0, t_f]$

$$v(\hat{\boldsymbol{\tau}}^{\hat{R}}) = \hat{u}^{\hat{R}}(t) = -\operatorname{sgn} \hat{\phi}^{\hat{R}}(t) = \operatorname{sgn} \langle p_0\tilde{\boldsymbol{\lambda}}(t), \tilde{\mathbf{b}}(\tilde{\mathbf{x}}(t))\rangle. \quad (3.45)$$

Consequently, it is concluded that if the optimal control problem considered has a nonsingular (i.e., bang–bang) solution, a solution obtained by the STVM satisfies the TPBVP derived from the maximum principle; and if the TPBVP has a unique solution, the solution obtained by the STVM is the unique optimal solution to the optimal control problem.

3.3.3 Computational Algorithm and Application

The switching-time-variation method for variable terminal time problems is summarized by the computational algorithm given in this section. This algorithm can be modified for problems with fixed terminal time merely by letting $\tau_{N+1}^N = t_f$ and $\delta t_f = 0$. It is assumed that the state equation and the adjoint equation are given by (3.21) and (C.2), respectively, along with initial state and appropriately adjoined cost functional.

Then the algorithm consists of the following steps.

1. Guess a number of switchings N and a switching vector $\boldsymbol{\tau}_i^N$.

2. Obtain the modified state equation (3.26) by substituting $v(\boldsymbol{\tau}_i^N)$ into the original state equation (3.3). Then, starting from $\tilde{\mathbf{x}}(t_0) = \tilde{\mathbf{x}}_0$, integrate (3.26) from t_0 to t_f; t_f can be obtained sequentially from the desired terminal condition, or guessed, for the first iteration if no stopping condition is given. Though the initial choice of the terminal time t_f is not usually critical, it is desirable to choose the value as close as possible to optimal in order to reduce computation time. Store the resulting state trajectory $\tilde{\mathbf{x}}_i(t)$, for $t \in [t_0, t_f]$, and the switching vector $\boldsymbol{\tau}_i^N$.

3. Compute the adjoint trajectory $\tilde{\boldsymbol{\lambda}}_i(t)$ by integrating (C.2) backward in time using $\tilde{\mathbf{x}}_i(t)$ from step 2 and the boundary condition

$$\tilde{\boldsymbol{\lambda}}_i(t_f) = \left.\frac{\partial I}{\partial \tilde{\mathbf{x}}}\right|_{\tilde{\mathbf{x}}_i(t_f)}.$$

Store the values of the adjoint vector $\tilde{\boldsymbol{\lambda}}_i(t)$ for $t = \tau_{ij}^N$ $(j = 1, \ldots, N)$.

4. Compute the gradient vector $-\boldsymbol{\phi}_i^N$ and compute $\delta\tau_{ij}^N = k_{ij}\phi_{ij}$ by choosing a reasonable k_{ij} ($j = 1, \ldots, N$). Note that if the k_{ij} are picked too small, the convergence of the solution is very slow; on the other hand, if the k_{ij} are picked too large, the condition $\boldsymbol{\tau}_i^N + \mathbf{K}_i\boldsymbol{\phi}_i \in T^N$ or $I(\boldsymbol{\tau}^N + \delta\boldsymbol{\tau}^N) = I(\boldsymbol{\tau}_i^N) + \Delta I_k$ may not be satisfied.

5. Compute a new switching vector $\boldsymbol{\tau}_{i+1}^N$ from $\boldsymbol{\tau}_{i+1}^N = \boldsymbol{\tau}_i^N + \delta\boldsymbol{\tau}_i^N$.

6. Repeat steps 2–5 until $\sum_{j=1}^N (\delta\tau_{ij}^N)^2 < \varepsilon$, where $\varepsilon > 0$ is a small given number.

7. Check if any of the intervals between the switching time approaches zero. If so, discard those switching times and renumber the switching times and the number of switchings.

8. Compute $\phi(t)$ and check if any of the zeros of $\phi(t)$ are between two switching times. If so, increase the number of switchings accordingly and redefine the switching vector from (B.2) and (B.3).

9. Repeat steps 2–8 until all the zeros of $\phi(t)$ coincide with the switching times.

10. Increase the number of switchings, and repeat steps 2–9 until the cost functional cannot be decreased by increasing the number of switchings.

11. Print the final switching times, the state trajectory, and the cost.

Example *Time-optimal control.* Moon and Mohler [29] show that the dynamics of a rotating searchlight may be approximated by

$$\begin{aligned} \dot{x}_1 &= x_2 \\ \dot{x}_2 &= -a_2(1 + u_2)x_2 + a_1u_1, \end{aligned} \tag{3.46}$$

where x_1 is angular position, $\mathbf{x}(t_0) = \mathbf{x}_0$, and $|u_i| \le 1$ ($i = 1, 2$). The problem is to control system (3.46) to the origin in minimal time with an admissible control. It follows from the maximum principle that extremal controls are

$$u_1(t) = \operatorname{sgn}\big(p_2(t)\big) \qquad \text{and} \qquad u_2(t) = \operatorname{sgn}\big(p_2(t)x_2(t)\big), \tag{3.47}$$

where $p_2(t)$ is a solution to the adjoint of (3.46).

It can be shown that the problem does not have a singular solution, that $p_2(t)$ has at most one zero, and that $p_2(t)$ has a maximum of two zeros [29]. Then Fig. 3.10 shows the switching times and cost as a function of iteration for the solution of this problem by the STVM. It is found approximately that $u = (-1, 1)$ on $(0, 1.2\text{ sec})$, $u = (-1, -1)$ on

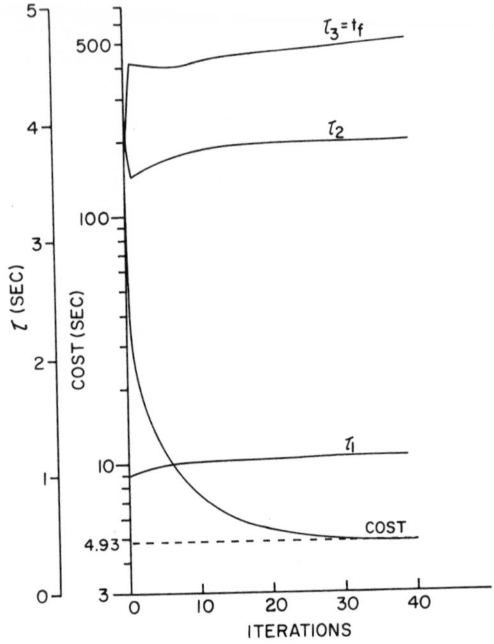

Fig. 3.10. Cost and switching times versus iteration for time-optimal policy from STVM.

(1.2, 3.8 sec) and $u = (+1, +1)$ on (3.8, 4.8 sec). The numerical computations were performed by an IBM 360/75J. Each run required about 4.05 sec and less than 64K bytes. Here $a_1 = 2$, $a_2 = 1$, $x_1(0) = 5$, $x_2(0) = 10$.

3.3.4 Control Rate Constraint

In this section, admissible controls are further constrained so that their magnitudes cannot be changed instantaneously. The switching-time-variation method is extended to solve the problem stated in conjunction with (3.21). For convenience the scalar control problem is reformulated as follows: Given the state equation

$$d\mathbf{x}/dt = \mathbf{a}(\mathbf{x}) + \mathbf{b}(\mathbf{x})u, \tag{3.48}$$

where $\mathbf{x}(t_0) = \mathbf{x}_0$, $|u| \leq 1$, and $|\dot{u}| \leq 1$, find an admissible control that transfers the system from $\mathbf{x}_0$ to the smooth terminal manifold θ_f such that

$$J = \int_{t_0}^{t_f} [a_0(\mathbf{x}) + b_0(\mathbf{x})u]\, dt \tag{3.49}$$

is minimized. Define

$$\begin{aligned} \mathbf{z} &= [x_a \; x_1 \cdots x_n \; u]^{\mathrm{T}} \\ &= [z_0 \cdots z_{n+1}]^{\mathrm{T}} \end{aligned} \tag{3.50}$$

$$\mathbf{h}(\mathbf{z}) = \begin{bmatrix} a_0(x) + b_0(\mathbf{x})u \\ \mathbf{a}(\mathbf{x}) + \mathbf{b}(\mathbf{x})u \end{bmatrix} \tag{3.51}$$

$$g(\mathbf{z}) = u^2 - 1 \tag{3.52}$$

and

$$w = \dot{u}, \tag{3.53}$$

where $\mathbf{z} \in R^{n+2}$, and scalar control w is constrained so that $|w| \leq 1$. Hence the process is transformed so that the state $\mathbf{y}$ is described by a nonlinear system

$$d\mathbf{z}/dt = \mathbf{h}(\mathbf{z}) + \mathbf{c}w, \tag{3.54}$$

where the $n + 2$ vector $\mathbf{c} = [0 \cdots 0 \;\; 1]^{\mathrm{T}}$. The the optimal control problem is transformed to $(n + 2)$-dimensional state space, and may be solved in principle at least by the restricted maximum principle [7, Chapter 6]. Again, the Hamiltonian is described by

$$H(\mathbf{z}, \mathbf{p}, \mathbf{w}) = \langle \mathbf{p}, [\mathbf{h}(\mathbf{z}) + \mathbf{c}w] \rangle, \tag{3.55}$$

where the costate $\mathbf{p}(t)$ satisfies the adjoint equation of (3.54) or

$$d\mathbf{p}/dt = [\partial \mathbf{h}/\partial \mathbf{z}]^{\mathrm{T}} \mathbf{p} \tag{3.56}$$

for $\mathbf{p}(t_f) = 0$ where $\mathbf{p} \in R^{n+2}$. Also, on the state constraint boundary described by $g(\mathbf{z}) = z_{n+1}^2 - 1 = 0$, it is seen that $\dot{g}(\mathbf{z}) = 2z_{n+1}\dot{z}_{n+1} = 0$ or

$$|\dot{z}_{n+1}| = |u| = 1$$

and

$$w = \dot{u} = 0. \tag{3.57}$$

Maximization of (3.55) yields

$$w = \operatorname{sgn}(p_{n+1}) \tag{3.58}$$

for $g(x) < 0$ or $|u| < 1$. Again, it is assumed that the optimal control can be described by a sequence of step functions so that

$$\begin{aligned} w(t) = & \sum_{j=0}^{e_1-1} (-1)^j[\delta_{-1}(t-\tau_j) - \delta_{-1}(t-\tau_{j+1})] \\ & - \sum_{e_1+1}^{e_2-1} (-1)^j[\delta_{-1}(t-\tau_j) - \delta_{-1}(t-\tau_{j+1})] \\ & + \sum_{j=e_2+1}^{e_3-1} (-1)^j[\delta_{-1}(t-\tau_j) - \delta_{-1}(t-\tau_{j+1})] \\ & - \cdots + \cdots \end{aligned} \tag{3.59}$$

where $\tau_0 = t_0$, τ_j $(j = 1, 2, \ldots)$ are the zeros of $p_{n+1}(t)$ and τ_{e_i} and τ_{e_i+1} $(i = 1, 2, \ldots)$ are the times when the state trajectory enters and leaves its constraint boundary for the ith time.

As in the previous section, switching vectors and a switching set may be defined so that the cost functional $J = z_0(t_f)$ depends continuously on a switching vector $\boldsymbol{\tau}^N$. Again, with $J = I(\boldsymbol{\tau}^N) = z_0(t_f)$, there exists an element $\boldsymbol{\tau}^{N*}$ in the switching set T^N so that

$$I^{N*} = \min_{\boldsymbol{\tau}^N \in T^N} I(\tau^N) = I(\boldsymbol{\tau}^{N*})$$

is monotone increasing and convergent to $I^* \geq J^0$. Also, there exists an integer N_0 such that $I^{N_0*} = I^{N*}$ for all $N \geq N_0$.

The computation of the gradient of the performance index I with respect to the switching vector, and the generation of a nonincreasing sequence of cost functionals may be computed as before. Again, the STVM satisfies the TPBVP derived from the restricted maximum principle. These conditions are given by (3.55) to (3.58) along with the so-called jump conditions on $\mathbf{p}(t)$ at the state constraint corners [7].

3.4 FUEL-OPTIMAL CONTROL

The optimal control policy computed in this section must drive the system (3.4) from an initial state $\mathbf{x}(t_0) = \mathbf{x}_0$, to the terminal state $\mathbf{x}(t_f) \in \theta_f$ with an admissible control $\mathbf{u}(t) \in U$, where

$$U = \{\mathbf{u} \mid u_k| \leq 1, k = 1, \ldots, m\},$$

so that

$$J = \int_{t_0}^{t_f} \left(e_0 + \sum_{k=1}^{m} e_k |u_k| \right) dt \tag{3.60}$$

is minimized. Here $e_0, \ldots, e_m$ are positive weighting constants.

Again the form of optimal control may be obtained from the maximum principle with

$$H(\mathbf{x}, \mathbf{p}, \mathbf{u}) = p_0 \left(e_0 + \sum_{k=1}^{m} e_k |u_k| \right) + \langle \mathbf{p}, \mathbf{a}(\mathbf{x}) + \mathbf{b}(\mathbf{x})\mathbf{u} \rangle \tag{3.61}$$

$$\frac{d\mathbf{p}}{dt} = -\frac{\partial}{\partial \mathbf{x}} [\mathbf{a}(\mathbf{x}) + \mathbf{b}(\mathbf{x})\mathbf{u}]^{\mathrm{T}} \mathbf{p}, \tag{3.62}$$

and p_0 is a nonpositive constant. In order to maximize (3.62), the fuel-optimal control is described by

$$u_k = \operatorname{dez}(\alpha_k), \qquad k = 1, \ldots, m, \tag{3.63}$$

where

$$\operatorname{dez}(\alpha_k) = \begin{cases} \operatorname{sgn}(\alpha_k) & \text{if } |\alpha_k| > 1 \\ 0 & \text{if } |\alpha_k| < 1, \end{cases} \tag{3.64}$$

and

$$\alpha_k = \frac{\sum_{j=1}^{n} b_{jk} p_j}{e_k |p_0|},$$

where b_{jk} is the jkth element of $\mathbf{B}$. Here it is assumed that $\alpha_k(t)$ does not vanish on any finite time interval.

3.4.1 Sequential Algorithm

In order to simplify the application of the switching-time-variation method, as in Section 3.3 it is assumed that the control is scalar, that an optimal control exists, and that the terminal state constraint is included by means of penalty terms added to (3.60) so that

$$J = \int_{t_0}^{t_f} [e_1 |u| + a_0(\mathbf{x}) + b_0(\mathbf{x}) u] \, dt, \tag{3.65}$$

where

$$a_0(\mathbf{x}) = e_0 + 2 \sum_{i=1}^{q} \rho_i \psi_i \langle \partial \psi / \partial \mathbf{x}, \mathbf{B}(\mathbf{x}) \rangle.$$

Here ρ_i is a positive constant, $i = 1, \ldots, q$, and ψ_i is the ith component of $\boldsymbol{\psi}(x) = \mathbf{0}$ in (3.16). (Convergence of the latter problem to the original problem as $\rho_i \to \infty$, $i = 1, \ldots, q$, is shown by Russell [28].)

Since

$$\mathrm{dez}(\alpha) = \tfrac{1}{2}[\mathrm{sgn}(\alpha - 1) + 1] + \tfrac{1}{2}[\mathrm{sgn}(\alpha + 1) - 1],$$

it follows that (3.63) can be expressed by

$$u(t) = \tfrac{1}{2}\left\{\sum_{j=1}^{N^+} [\delta_{-1}(t - \tau_{2j-1}^+) - \delta_{-1}(t - \tau_{2j}^+)] - \sum_{j=1}^{N^-} [\delta_{-1}(t - \tau_{2j-1}^-) - \delta_{-1}(t - \tau_{2j}^-)]\right\}, \tag{3.66}$$

where τ_j^+, $j = 0, \ldots, 2N^+$, are the zeros of $\alpha(t) - 1$, and τ_j^-, $j = 0, \ldots, 2N^-$, are the zeros of $\alpha(t) + 1$. Here

$$\alpha(t) = \langle \mathbf{b}, \mathbf{p}(t) \rangle, \tag{3.67}$$

for scalar control with $\mathbf{B}$ replaced by $\mathbf{b}$ and $\mathbf{u}$ by u in (3.4).

Now, the numbers of switchings N^+ and N^- are computed as shown in Appendix D, and the control may be specified by the switching vectors of $(2N^+) + (2N^-)$ dimension,

$$\boldsymbol{\tau}^N = \begin{bmatrix} \boldsymbol{\tau}^{N+} \\ \boldsymbol{\tau}^{N-} \end{bmatrix}, \tag{3.68}$$

where

$$\boldsymbol{\tau}^{N+} = \begin{bmatrix} \tau_1^{N+} \\ \vdots \\ \tau_{2N^+}^{N+} \end{bmatrix}, \qquad \boldsymbol{\tau}^{N-} = \begin{bmatrix} \tau_1^{N-} \\ \vdots \\ \tau_{2N^-}^{N-} \end{bmatrix}.$$

The switching set T^N is defined by

$$T^N = \{\boldsymbol{\tau}^N \mid \boldsymbol{\tau}^{N+} \in T^{N+} \text{ and } \boldsymbol{\tau}^{N-} \in T^{N-}\}, \tag{3.69}$$

where

$$T^{N+} = \{\boldsymbol{\tau}^{N+} \mid t_0 \leq \tau_1^{N+} \leq \cdots \leq \tau_{2N^+}^{N+} \leq t_f\}$$

$$T^{N-} = \{\boldsymbol{\tau}^{N-} \mid t_0 \leq \tau_1^{N-} \leq \cdots \leq \tau_{2N^-}^{N-} \leq t_f\}.$$

A control

$$v = v(\boldsymbol{\tau}^N) = v(\tau^{N+}) + v(\tau^{N-}), \tag{3.70}$$

where

$$v^+ = v(\boldsymbol{\tau}^{N+}) = \sum_{j=1}^{N+} [\delta_{-1}(t - \tau_{2j-1}^{N+}) - \delta_{-1}(t - \tau_{2j}^{N+})]$$

$$v^- = v(\boldsymbol{\tau}^{N-}) = -\sum_{j=1}^{N-} [\delta_{-1}(t - \tau_{2j-1}^{N-}) - \delta_{-1}(t - \tau_{2j}^{N-})],$$

and a control set

$$V^N = \{v(\boldsymbol{\tau}^N) \mid \boldsymbol{\tau}^N \in T^N, N = 1, 2, \ldots\}$$

are utilized for $u(t)$ and U in (3.4). Then the system equations are

$$\begin{aligned} d\tilde{\mathbf{x}}/dt &= \tilde{\mathbf{a}}(\tilde{\mathbf{x}}) && \text{for} \quad t \in \{t \mid v = 0\} \\ d\tilde{\mathbf{x}}/dt &= \tilde{\mathbf{a}}(\tilde{\mathbf{x}}) + \tilde{\mathbf{b}}(\tilde{\mathbf{x}}) + \mathbf{e} && \text{for} \quad t \in \{t \mid v = +1\} \\ d\tilde{\mathbf{x}}/dt &= \tilde{\mathbf{a}}(\tilde{\mathbf{x}}) - \tilde{\mathbf{b}}(\tilde{\mathbf{x}}) + \mathbf{e} && \text{for} \quad t \in \{t \mid v = -1\}, \end{aligned} \tag{3.71}$$

where $\tilde{\mathbf{x}}$, $\tilde{\mathbf{a}}(\tilde{\mathbf{x}})$, and $\tilde{\mathbf{b}}(\tilde{\mathbf{x}})$ are defined with (3.17), and the

$$\mathbf{e} = \begin{bmatrix} e_1 \\ 0 \\ \vdots \\ 0 \end{bmatrix}$$

with $\mathbf{e} \in R^{n+1}$.

Since (3.4) is well defined on each interval with a unique continuous solution which depends continuously on $\mathbf{x}_0$, the cost, $J = x_a(t_f)$, depends continuously on the switching vector $\boldsymbol{\tau}^N$. As before, let

$$I(\boldsymbol{\tau}^N) = x_a(t_f) = J(v(\boldsymbol{\tau}^N)). \tag{3.72}$$

Then there exists an I^N on the range of I and a $\boldsymbol{\tau}^{N*}$ in T^N such that

$$I^{N*} = \min_{\boldsymbol{\tau}^N \in T^N} I(\boldsymbol{\tau}^N) = I(\boldsymbol{\tau}^{N*}). \tag{3.73}$$

T^N is a compact set in R^n, and $T^N \subset T^{N+1}$ so that

$$I^{N*} = \min_{T^N} I(\boldsymbol{\tau}^N) \geq \min_{T^{N+1}} I(\boldsymbol{\tau}^{N+1}) = I^{N+1*}. \tag{3.74}$$

Hence the sequence $\{I^{N*}\}$ is nonincreasing with a lower bound specified by the optimal cost J^0, and $\{I^{N*}\}$ converges to some $I^* \geq J^0$ such that there exists some integer N_0 with

$$I^{N*} = I^{N_0*} \geq J^0$$

for all $N \geq N_0$. Consequently, a gradient algorithm such as the STVM can be used to solve the TPBVP associated with Eqs. (3.14), (3.62), and (3.63).

3.4.2 Computation of Switching Vectors

A successive approximation to a switching vector, which corresponds to a local minimum cost, is computed with a gradient method. To obtain the gradient it is convenient to start with the variational state equation

$$\frac{d\tilde{\mathbf{x}}}{dt} = \frac{\partial}{\partial \tilde{\mathbf{x}}}[\tilde{\mathbf{a}}(\tilde{\mathbf{x}}) + \tilde{\mathbf{b}}(\tilde{\mathbf{x}})v]\delta\tilde{\mathbf{x}} + \tilde{\mathbf{b}}(\tilde{\mathbf{x}})\delta v + \mathbf{e}\delta\,|v|$$

and adjoint system

$$\frac{d\tilde{\boldsymbol{\lambda}}}{dt} = -\frac{\partial}{\partial \tilde{\mathbf{x}}}[\tilde{\mathbf{a}}(\tilde{\mathbf{x}}) + \tilde{\mathbf{b}}(\tilde{\mathbf{x}})v]^{\mathrm{T}}\tilde{\boldsymbol{\lambda}}, \tag{3.75}$$

where

$$\tilde{\boldsymbol{\lambda}}(t_f) = \left.\frac{\partial J}{\partial \tilde{\mathbf{x}}}\right|_{t_f}.$$

Then the variation in cost may be computed by

$$\Delta J = \langle \partial J/\partial \tilde{\mathbf{x}}, \delta\tilde{\mathbf{x}}\rangle|_{t_f} = \int_{t_0}^{t_f} [\langle \tilde{\boldsymbol{\lambda}}, \tilde{\mathbf{b}}(\tilde{\mathbf{x}})\rangle \delta v + \langle \tilde{\boldsymbol{\lambda}}, \mathbf{e}\rangle \delta|v|]\, dt,$$

where $\delta v = \delta v^+ + \delta v^-$, $\delta|v| = \delta v^+ - \delta v^-$,

$$\delta v^+ = -\sum_{j=1}^{N+} [\Delta(t - \tau_{2j-1}^+)\delta\tau_{2j-1}^+ - \Delta(t - \tau_{2j}^+)\delta\tau_{2j}^+]$$

$$\delta v^- = \sum_{j=1}^{N-} [\Delta(t - \tau_{2j-1}^-)\delta\tau_{2j-1}^- - \Delta(t - \tau_{2j}^-)\delta\tau_{2j}^-],$$

and n superscripts are dropped for convenience. Substitution and condensation of terms yield

$$\Delta J = -\langle \boldsymbol{\phi}, \delta\boldsymbol{\tau}\rangle, \tag{3.76}$$

where $\boldsymbol{\phi} = [\phi_1, \ldots, \phi_N]^{\mathrm{T}}$,

$$\phi_j = (-1)^{j-1}e_1(\beta_j{}^+ + 1), \qquad j = 1, \ldots, 2N^+,$$

$$\phi_{j+2N} = (-1)^j e_1(\beta_j{}^- - 1), \qquad j = 1, \ldots, 2N^-,$$

$$\beta_j^+ = \frac{\langle \tilde{\boldsymbol{\lambda}}(t), \tilde{\mathbf{b}}(\tilde{\mathbf{x}}(t))\rangle}{e_1}\bigg|_{t=\tau_j^+}, \qquad j = 1, \ldots, 2N^+,$$

$$\beta_j^- = \frac{\langle \tilde{\boldsymbol{\lambda}}(t), \tilde{\mathbf{b}}(\tilde{\mathbf{x}}(t))\rangle}{e_1}\bigg|_{t=\tau_j^-}, \qquad j = 1, \ldots, 2N^-.$$

The gradient of I with respect to the switching vector is given by

$$\text{grad } I(\boldsymbol{\tau}^N) = -\boldsymbol{\phi}, \tag{3.77}$$

and the nonincreasing sequence of cost functionals and corresponding switching vectors can be computed from

$$\boldsymbol{\tau}_{i+1} = \boldsymbol{\tau}_i + \mathbf{K}_i\boldsymbol{\phi}_i, \tag{3.78}$$

where $\boldsymbol{\tau}_i$ is the current switching vector, $\boldsymbol{\tau}_{i+1}$ is the improved switching vector, $\boldsymbol{\phi}_i$ is the negative of the gradient vector, and $\mathbf{K}_i$ is an $(N \times N)$-dimensional diagonal matrix with sufficiently small nonnegative diagonal elements such that $\boldsymbol{\tau}_i + \mathbf{K}_i\boldsymbol{\phi}_i \in T^N$ and $I(\boldsymbol{\tau}_i + \mathbf{K}_i\boldsymbol{\phi}_i) \sim I(\boldsymbol{\tau}_i) - \langle \boldsymbol{\phi}_i, \mathbf{K}_i\boldsymbol{\phi}_i\rangle$.

Since the nonincreasing sequence $\{I_i\}$ is bounded below by I^{N*}, the sequence converges to some $\hat{I}^N \geq I^{N*}$. Also, since I is continuous and T^N is compact, there exists a $\hat{\boldsymbol{\tau}} \in T^N$ with $\hat{I}^N = T(\hat{\tau})$ and

$$\mathbf{K}\boldsymbol{\phi}(\hat{\tau}) = \mathbf{0}. \tag{3.79}$$

3.4.3 Application

Fuel-optimal control of the automobile model given in Section 1.2 by (1.8) is discussed in this subsection.

It is assumed for convenience that the admissible set U is defined by $|u_1| \leq 1$ and $|u_2| \leq 1$; let $a = -kc_f/m = -kc_b/m$. It is a simple matter, however to let $u_1' = 2u_1 - 1$ with $|u_1'| \leq 1$, so that physical constraints of the form $0 \leq u \leq 1$ are symmetrical.

The problem considered here is to drive the system (1.8) from an initial state $\mathbf{x}(t_0) = \mathbf{x}_0$ to the origin in the state space so that

$$J = \int_{t_0}^{t_f} (1 + e_1|u_2|)\, dt, \qquad \text{where} \quad e_1 > 0, \tag{3.80}$$

is a minimum. The Hamiltonian and the costate equation for this problem are

$$H = p_0(1 + e_1|u_2|) + p_1x_2 + p_2[ax_2(1 + u_1) + u_2] \tag{3.81}$$

and

$$\begin{aligned} \dot{p}_0 &= \dot{p}_1 = 0 \\ \dot{p}_2 &= -p_1 - a(1 + u_1)p_2. \end{aligned} \tag{3.82}$$

From (3.81) and the maximum principle,

$$u_1 = \operatorname{sgn}(s_1) \qquad \text{if} \quad s_1 \neq 0 \tag{3.83}$$

and

$$u_2 = \operatorname{dez}(s_2) \qquad \text{if} \quad |s_2| \neq 1, \tag{3.84}$$

where u_1 and u_2 are extremal controls, and

$$s_1(t) = ap_2(t)x_2(t) \tag{3.85}$$

and

$$s_2(t) = \frac{p_2(t)}{e_1|p_0|}. \tag{3.86}$$

It is shown now that this fuel-optimal problem does not have a singular solution, except along the trivial trajectory given by $x_1 = x_2 = 0$.

For u_1 to be singular, s_1 must vanish on some nonzero interval. That is, either $p_2(t)$ or $x_2(t)$ must be zero on a nonzero interval.

Suppose $x_2 = 0$. Then it follows that $\dot{x}_1 = \dot{x}_2 = 0$, which implies that the system is in equilibrium state. Therefore, assume that $x_2 \neq 0$. Then $p_2(t)$ must be zero for a singular solution. If $p_2 = 0$, then $\dot{p}_2$ must be zero also, and hence

$$\dot{p}_2 = -p_1 - a(1 + u_1)p_2 = -p_1 = 0.$$

But this is impossible according to the maximum principle. Therefore, optimal u_1 must be bang–bang, except along the trivial trajectory $x_1 = x_2 = 0$.

Now, in order for u_2 to be singular, $|s_2(t)| = 1$ must be satisfied on some nonzero interval. That is, either $s_2 = -1$ or $s_2 = +1$. Suppose $s_2 = -1$. Then, from (3.86),

$$p_2 = e_1 p_0 = \text{constant}. \tag{3.87}$$

Hence, $\dot{p}_2 = -p_1 - a(1 + u_1)p_2 = 0$, and

$$p_1 = -a(1 + u_1)p_2 = -ae_1p_0(1 + u_1). \tag{3.88}$$

Substitution of Eqs. (3.87) and (3.88) into (3.81) yields

$$H = p_0 + p_0e_1|u_2| + e_1p_0u_2. \tag{3.89}$$

Then, from $s_2(t) = -1$ and from (3.89), it can be seen that $-1 \leq u_2 \leq 0$, and that

$$H = p_0 - p_0 e_1 u_2 + p_0 e_1 u_2 = p_0 < 0.$$

But, according to the maximum principle, the extremal Hamiltonian H for a free terminal time problem must be zero. Hence, the switching function, $s_2(t) = -1$, cannot give an extremal singular solution.

For $s_2(t) = +1$ it is seen that

$$H = p_0 + p_0 e_1(|u_2| - u_2). \tag{3.90}$$

But, since $s_2(t) = +1$, it follows that $0 \leq u_2 \leq 1$ and that $H = p_0 < 0$. Hence the switching function $s_2(t) = +1$ cannot be an extremal switching function either. Hence, a nontrivial singular solution does not exist.

It is readily seen that $x_2(t)$ and $p_2(t)$ each can have at most one zero on (t_0, t_f) and that $p_2(t)$ is monotonic. Therefore, $s_2 = ax_2p_2$ can have at most two zeros on (t_0, t_f), and $s_2 = p_2(t)/e_1|p_0|$ is monotonic and can have at most two switchings. Also, it can be seen from $s_1 = ax_2p_2$ and $s_2 = p_2/e_1|p_0|$ that the zero of s_1 which corresponds to $p_2 = 0$ must lie between the two switchings of s_2. That is, if $p_2(t_c) = 0$, $|s_1(t_b)| = |s_1(t_d)| = 1$, and $t_b < t_d$, then the condition

$$t_b < t_c < t_d \tag{3.91}$$

must hold. Hence the maximum number of optimum switchings for the fuel-optimal control example is four.

At this point it can be assumed that the forms of extremal controls and state trajectories are given since the solution of (1.8) is well known for such piecewise constant controls. Now, the exact locations of the switching times are computed by using the switching-time-variation method. For numerical computations, assume that $t_0 = 0$, $a = 1$, $\rho = 10$, $x_{10} = 5$, $x_{20} = 10$, and $e_1 = 1$.

For the numerical values given above, it can be shown that the optimum number of switchings is four, and that the extremal control is given by

$$\begin{aligned} u_1(t) &= \delta_{-1}(t - t_0) - 2\delta_{-1}(t - \tau_1) + 2\delta_{-1}(t - \tau_3) - \delta_{-1}(t - t_f) \\ u_2(t) &= -\delta_{-1}(t - t_0) + \delta_{-1}(t - \tau_2) + \delta_{-1}(t - \tau_4) - \delta_{-1}(t - t_f), \end{aligned} \tag{3.92}$$

where $\delta_{-1}(t)$ is the unit step function, $\tau_1 = t_a$, $\tau_2 = t_b$, $\tau_3 = t_c$, and $\tau_4 = t_d$.

TABLE 3.1

NUMERICAL RESULTS FOR FUEL-OPTIMAL CONTROL

ITERATION NUMBER 0						
TAU	0.0	0.1000E 01	0.2000E 01	0.3000E 01	0.4000E 01	0.5000E 01
X1(T)	0.5000E 01	0.8756E 01	0.8244E 01	0.6733E 01	0.6080E 01	0.6559E 01
X2(T)	0.1000E 02	0.4887E 00	−0.1511E 01	−0.1511E 01	0.2045E 00	0.8370E 00
COST =	0.2266E 03					
PHI(T)	0.0	0.7991E 02	0.3907E 03	0.4889E 02	0.1190E 03	−0.5963E 02
ITERATION NUMBER 10						
X1(T)	0.5000E 0.	0.8801E 01	0.4210E 01	0.3534E 01	0.1664E 01	0.1609E 01
X2(T)	0.1000E 02	0.4346E −01	−0.4285E 01	−0.4285E 01	−0.5453E 00	0.1121E −01
COST =	0.2126E 02					
PHI(T)	0.0	0.1945E 01	0.3875E 02	0.3170E 02	0.1287E 02	−0.2402E 01
ITERATION NUMBER 20						
TAU	0.0	0.1187E 01	0.3566E 01	0.3698E 01	0.4571E 01	0.4776E 01
X1(T)	0.5000E 01	0.8801E 01	0.3197E 01	0.2573E 01	0.6190E 00	0.5169E 00
X2(T)	0.1000E 02	0.2355E −01	−0.4735E 01	−0.4735E 01	1.8262E 00	−0.2116E 00
COST =	0.1011E 02					
PHI(T)	0.0	0.3536E 00	0.9881E 01	0.9659E 01	0.1137E 01	0.4222E 01

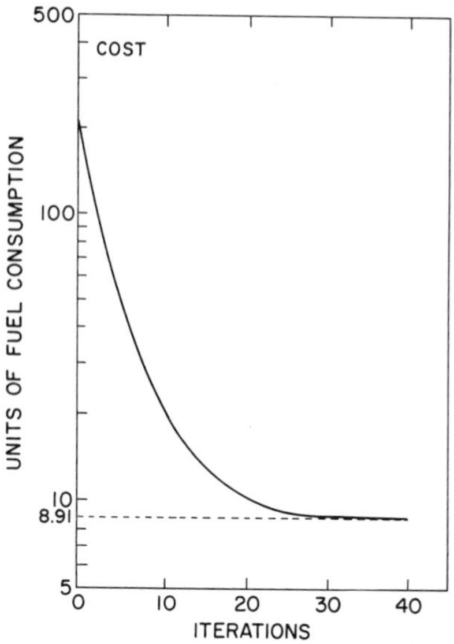

Fig. 3.11. Cost versus iteration for fuel-optimal policy from STVM.

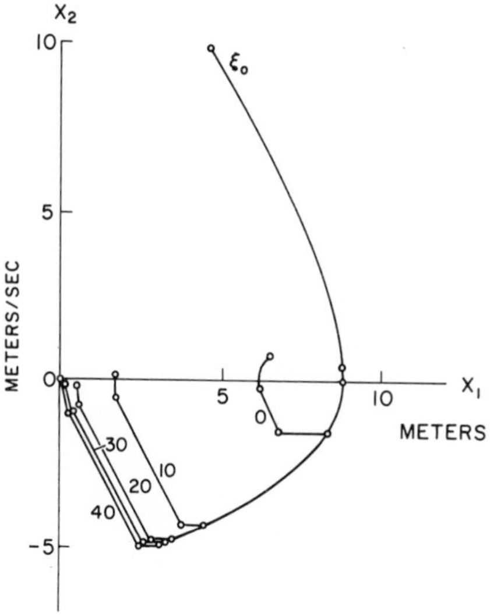

Fig. 3.12. Fuel-optimal phase trajectories from STVM. The numbers along the trajectories refer to the number of iterations.

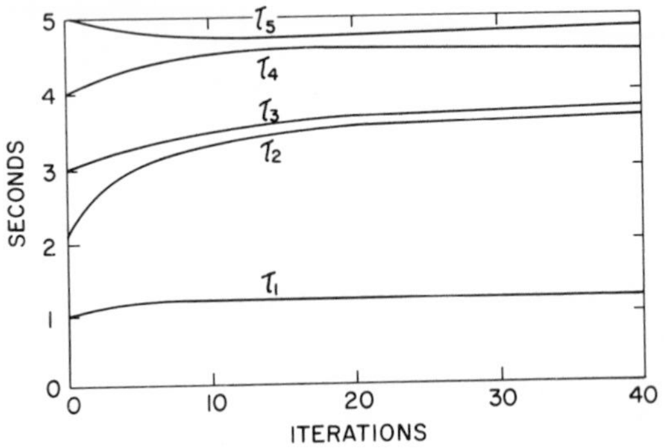

Fig. 3.13. Fuel-optimal switching times versus iteration from STVM.

To start the numerical computations, it is initially assumed that $\boldsymbol{\tau}_0 = (1.0, 2.0, 3.0, 4.0, 5.0)^T$, where $\tau_5 = t_f = 5.0$.

Application of the STVM, with the initial value and the control given above, shows that the sequence of the switching vector $\{\boldsymbol{\tau}_i\}$ converges to $(1.2, 3.64, 3.78, 4.54, 4.83)^T$. The final cost corresponding to this switching vector is 8.91.

The numerically computed results are shown in Table 3.1 and Figs. 3.11 to 3.13.

The numerical computation for this fuel-optimal example was carried out by an IBM 360/75J, The computation time was approximately 3.85 sec, and the total computer memory required was less than 64K bytes.

3.5 DISCRETE BILINEAR CONTROL

Bang–bang-controlled bilinear processes are a subclass of another interesting class of bilinear systems, namely, those with piecewise constant control. Such models may arise, for example, when a bilinear control process involves a stepping motor. Control of a nuclear reactor (a bilinear fission process) with neutron absorbers or reflectors driven by a stepping motor is one application of such a model.

Piecewise constant control may be envisioned as discrete control which is passed through a so-called zero-order hold or boxcar filter [30]. Again in the case of linear plants, discrete control processes or sampled data systems have been thoroughly studied in the literature. For example, Cadzow and Martens [30] provide a very adequate introduction. Since bilinear systems with piecewise constant control are really piecewise time-invariant linear systems, much of the theory for linear sampled data systems and even for the switching-time-variation method discussed in this chapter may be utilized. Rather than make an exhaustive detailed analysis, however, only a cursory "broad-brush" discussion is provided here. Such presentations are made in various sections with the hope to stimulate further investigation on the reader's part.

3.5.1 Piecewise Constant Control

Consider the bilinear process described by Eq. (3.3) with piecewise constant control

$$\mathbf{u}(t) = \sum_{j=0}^{\infty} \mathbf{u}(jT)[\delta_{-1}(t - jT) - \delta_{-1}(t - (j + 1)T)], \quad (3.93)$$

where T is the time between samples and again $\delta_{-1}(\cdot)$ is the unit step function. For convenience, let

$$\mathbf{u}_j = \mathbf{u}(jT)$$

and

$$\Delta_j = \delta_{-1}(t - jT) - \delta_{-1}(t - (j + 1)T),$$

so that (3.3) becomes

$$d\mathbf{x}/dt = \left[\mathbf{A} + \sum_{k=1}^{m} \sum_{j=0}^{\infty} \mathbf{B}_k u_{kj} \Delta_j\right]\mathbf{x} + \mathbf{C} \sum_{j=0}^{\infty} \mathbf{u}_j \Delta_j, \quad (3.94)$$

where u_{kj} is the kth element of control sample $\mathbf{u}_j$.

Obviously, (3.94) describes the dynamics of a piecewise, constant coefficient, linear system. The usual state matrix over the jth interval is given by

$$\mathbf{A} = \mathbf{A} + \sum_{k=1}^{m} \mathbf{B}_k u_{kj} \Delta_j. \quad (3.95)$$

A recursive solution to (3.95), with $\mathbf{x}(jT)$ already computed, is given by

$$\mathbf{x}(t) = \boldsymbol{\phi}_j(t - jT)\mathbf{x}_j + \int_{jT}^{t} \boldsymbol{\phi}_j(t - j\tau)\mathbf{C}\mathbf{u}_j\Delta_j \, d\tau, \tag{3.96}$$

where $\mathbf{x}_j = \mathbf{x}(jT)$, the state transition matrix, is given by

$$\boldsymbol{\phi}_j(t - jT) = \exp \mathbf{A}_j(t - jT),$$

and $t \in [jT, (j + 1)T)$. Consequently, computational methods utilized for linear systems and the switching-time-variation method, at least in principle, may be used to obtain optimal control policies of the class described by (3.93).

3.5.2 Optimal Control Application

Another form of discrete bilinear system is defined by

$$\mathbf{x}_{j+1} = \mathbf{A}\mathbf{x}_j + \sum_{k=1}^{m} \mathbf{B}_k u_{kj}\mathbf{x}_j + \mathbf{C}\mathbf{u}_j. \tag{3.97}$$

Optimal control policies for (3.97) may be computed by numerous programming methods or sometimes by the discrete maximum principle [31].

For example, suppose the bilinear system is described by phase variables ($x_1 = x, x_2 = \dot{x}_1, \ldots, x_n = \dot{x}_{n-1}$) with a zero $\mathbf{C}$ matrix and with scalar control so that

$$\mathbf{x}_{j+1} = \mathbf{A}\mathbf{x}_j + \mathbf{B}u_j\mathbf{x}_j, \tag{3.98}$$

where $\mathbf{x}(0) = \mathbf{x}_0$, and $\mathbf{A}$ and $\mathbf{B}$ are the usual matrices for n phase variables with the corresponding last rows defined by $\mathbf{a}^{\mathrm{T}}$ and $\mathbf{b}^{\mathrm{T}}$ with

$$\mathbf{a} = \begin{bmatrix} a_1 \\ \vdots \\ a_n \end{bmatrix} \quad \text{and} \quad \mathbf{b} = \begin{bmatrix} b_1 \\ \vdots \\ b_n \end{bmatrix}.$$

The problem is to find the control sequence $\{u_0, \ldots, u_{N-1}\}$ which drives (3.98) to $\mathbf{x}_N = \mathbf{0}$ so that

$$J = \sum_{j=0}^{N-1} \mathbf{x}_j^{\mathrm{T}}\mathbf{Q}\mathbf{x}_j \tag{3.99}$$

is a minimum, and so that

$$|u_j| \leq 1, \qquad j = 0, 1, \ldots, N - 1. \tag{3.100}$$

Here $\mathbf{Q}$ is an $n \times n$ positive-definite symmetric matrix. Again, the optimal control must maximize a scalar function defined by

$$H(\mathbf{x}_j, \mathbf{p}_{j+1}, u_j) = \mathbf{p}_{j+1}^{\mathrm{T}}[\mathbf{A}\mathbf{x}_j + \mathbf{B}\mathbf{x}_j u_j] + p_0 \mathbf{x}_j^{\mathrm{T}} \mathbf{Q} \mathbf{x}_j. \tag{3.101}$$

Hence the optimal policy has the form

$$u_j = -1 \qquad \text{for} \quad s = p_{j+1}^n(\mathbf{b}^{\mathrm{T}}\mathbf{x}_j) < 0$$

and

$$u_j = +1 \qquad \text{for} \quad s > 0. \tag{3.102}$$

Again the control $u(t)$ is said to be singular for $s = 0$, in which case the H does not depend on the control u_j. Similar to the continuous case, state and costate may be related through H by

$$\mathbf{x}_{j+1} = \partial H(\mathbf{x}_j, \mathbf{p}_{j+1}, u_j)/\partial \mathbf{x}_j, \qquad k = 0, 1, \ldots, N - 1 \tag{3.103}$$

and

$$\mathbf{p}_j = \partial H(\mathbf{x}_j, \mathbf{p}_{j+1}, u_j)/\partial \mathbf{x}_j, \qquad k = 1, \ldots, N. \tag{3.104}$$

Again p_0 must be a nonpositive constant. Elliott and Tarn [32] compute singular surfaces in state space from the condition that $s(\mathbf{x}, \mathbf{p}) = 0$. In other words, that

$$\mathbf{b}^{\mathrm{T}}\mathbf{x}_j = 0 \tag{3.105}$$

and

$$\mathbf{x}_{j+1} = \mathbf{A}\mathbf{x}_j, \tag{3.106}$$

or that $\mathbf{p}_j$ given by (3.104) is zero. It is readily seen that (3.105) and (3.106) are equivalent to the condition that the vectors

$$\mathbf{b}, \quad \mathbf{A}^{\mathrm{T}}\mathbf{b}, \quad \ldots, \quad \mathbf{A}^{n-1^{\mathrm{T}}}\mathbf{b}$$

are linearly independent [32].

3.6 QUASI-LINEAR PROGRAMMING

Quasi-linearization may be utilized to solve the identification problem as noted in Chapter I. Also, it is a powerful tool for the solution of optimal control problems for nonlinear systems. An excellent discussion of the method is provided by Bellman and Kalaba [33]. Solution to the problem is obtained by solving an approximating sequence of linearized system problems.

Another sequential solution that utilizes linearized optimization problems in conjunction with linear programming is discussed by Mohler and Shen [34] and by Mohler *et al.* [35]. This so-called quasi-

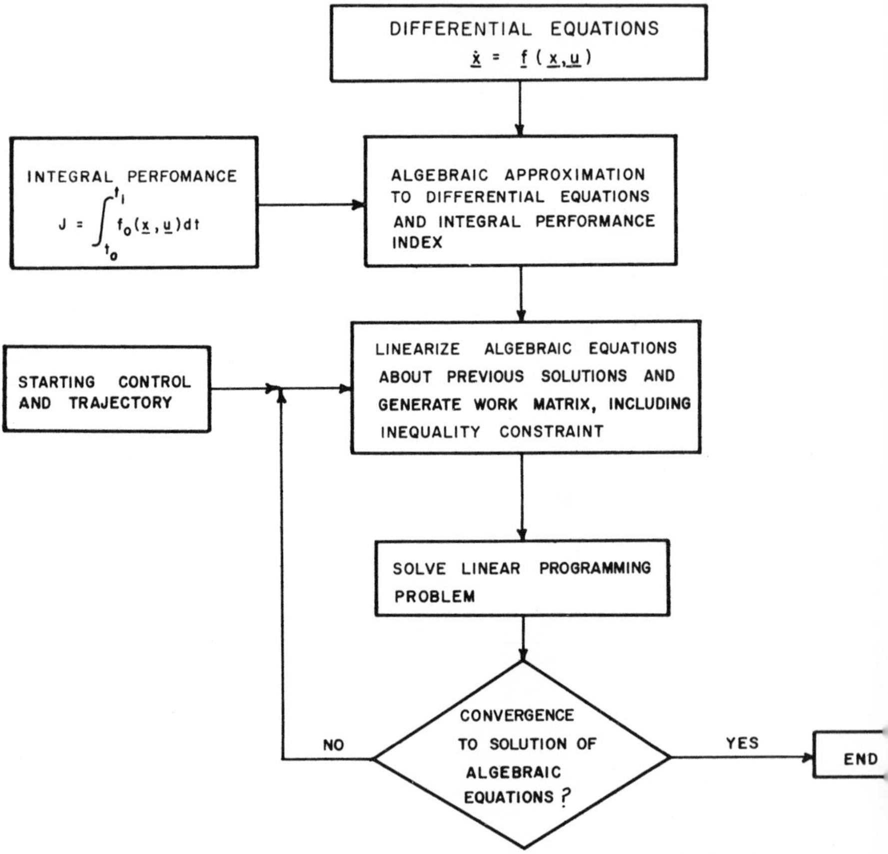

Fig. 3.14. Flow chart for quasi-linear programming algorithm [6].

linear programming algorithm is described by Fig. 3.14. The first step is to approximate the state differential equations,

$$d\mathbf{x}/dt = \mathbf{f}(\mathbf{x}, \mathbf{u}), \tag{3.107}$$

by difference equations such as

$$\mathbf{x}_{j+1} - \mathbf{x}_j = \tfrac{1}{2}\Delta t_j[\mathbf{f}(\mathbf{x}_j, \mathbf{u}_j) + \mathbf{f}(\mathbf{x}_{j+1}, u_{j+1})]. \tag{3.108}$$

The method is appropriate for many nonlinear processes, but particularly in the case of bilinear processes, which utilize divergent control modes, quasi-linear programming has been relatively efficient. For example, gradient schemes and computational methods such as the maximum principle which require solution to two-point boundary value problems frequently are not effective in such cases due to instabilities and accumulated computational errors. Also, it should be noted that quasi-linear programming is relatively effective for large-order systems (for which dynamic programming has trouble) and for problems with state constraints (for which maximum principle algorithms have trouble).

In order to utilize linear programming algorithms such as those that are available as canned problems with many computers (e.g., MPS/360), the difference equation is linearized either about the numerical integration solution or about the linear programming solution obtained in the previous iteration. Therefore, the approximating discrete linear system in the vicinity of $\mathbf{x}_j{}^*$ and $\mathbf{u}_j{}^*$ is

$$\begin{aligned}\mathbf{x}_{j+1} - \mathbf{x}_j = \frac{\Delta t_j}{2}\Bigg\{&\mathbf{f}(\mathbf{x}_j{}^*, \mathbf{u}_j{}^*) + \mathbf{f}(\mathbf{x}_j{}^*\mathbf{u}_{j+1}^*)\\ &+ \left[\frac{\partial \mathbf{f}}{\partial \mathbf{x}_j}\bigg|_*(\mathbf{x} - \mathbf{x}_j{}^*) + \frac{\partial \mathbf{f}}{\partial \mathbf{x}_{j+1}}\bigg|_*(\mathbf{x}_{j+1} - \mathbf{x}_{j+1}^*)\right]\\ &+ \left[\frac{\partial \mathbf{f}}{\partial \mathbf{u}_j}\bigg|_*(\mathbf{u}_j - \mathbf{u}_j{}^*) + \frac{\partial \mathbf{f}}{\partial \mathbf{u}_{j+1}}\bigg|_*(\mathbf{u}_{j+1} - \mathbf{u}_{j+1}^*)\right]\Bigg\}.\end{aligned} \tag{3.109}$$

Here it is assumed that Δt_j is fixed, and usually $\Delta t_j = \Delta t$ is a constant time interval. For fixed terminal time t_f, with $t_f = \sum_j \Delta t_j$, such an assumption is valid. For free terminal time with $\Delta t_i = \Delta t$, linearization in state-control-time space yields

$$\begin{aligned} \mathbf{x}_{j+1} - \mathbf{x}_j &= \frac{\Delta t^*}{2} [\mathbf{f}(\mathbf{x}_j^*, \mathbf{u}_j^*) + \mathbf{f}(\mathbf{x}_{j+1}^*, \mathbf{u}_{j+1}^*)] \\ &+ \frac{\Delta t^*}{2} \left[\frac{\partial \mathbf{f}}{\partial \mathbf{x}_j}\bigg|_* (\mathbf{x}_j - \mathbf{x}_j^*) + \frac{\partial \mathbf{f}}{\partial \mathbf{x}_{j+1}}\bigg|_* (\mathbf{x}_{j+1} - \mathbf{x}_{j+1}^*) \right] \\ &+ \frac{\Delta t^*}{2} \left[\frac{\partial \mathbf{f}}{\partial \mathbf{u}_j}\bigg|_* (\mathbf{u}_j - \mathbf{u}_j^*) + \frac{\partial \mathbf{f}}{\partial \mathbf{u}_{j+1}}\bigg|_* (\mathbf{u}_{j+1} - \mathbf{u}_{j+1}^*) \right] \\ &+ \tfrac{1}{2}[\mathbf{f}(\mathbf{x}_j^*, \mathbf{u}_j^*) + \mathbf{f}(\mathbf{x}_{j+1}^*, \mathbf{u}_{j+1}^*)](\Delta t - \Delta t^*). \end{aligned} \tag{3.110}$$

The integral performance index, Eq. (3.1), also must be expressed algebraically in order to solve the optimal control problem with linear programming. If another constraint equation is allowed, the performance index can be managed by an addended state variable x_0 with $J = x_0(t_f)$ and $dx_0/dt = f_0(\mathbf{x}, \mathbf{u})$ as done with the maximum principle. Linearization can be used as before. Computation is usually more efficient, however, if a simple algebraic approximation, such as

$$\int_{t_0}^{t_f} f_0(\mathbf{x}, \mathbf{u})\, dt = \sum_j \Delta t f_0(\mathbf{x}_j, \mathbf{u}_j),$$

is used.

Now the optimal control problem is in a form amenable to solution by the simplex algorithm. Geometrically, the linearized state and control constraints, including the reworked state equations, form an admissible region of solutions in state-control-time space. The simplex algorithm merely provides a search along edges of the region from one corner to another so that J decreases most. The process is repeated until J cannot be decreased further. Gass [36] provides a good introduction to linear programming. For more detailed and elegant presentations of the subject the reader is referred to Dantzig [37] or Hadley [38].

Solution of optimal control problems by linear programming was suggested some time ago [39], but its use has not been widely adopted. Rosen [40] proves convergence of a linear programming algorithm to the solution of certain optimal control problems with state equations that are convex functions of state and control. The system of difference equations is introduced by a penalty function addended to the original performance index. While convergence cannot generally be assured, it is achieved fortunately for the nuclear reactor control problem considered in Chapter IV.

The utility of quasi-linear programming is enhanced by availability of linear programming systems with modern digital computers. For example, the computations presented in Chapter IV were made by means of the Mathematical Programming System/360 (MPS/360). The algorithm consists of a sequence of linear programming solutions by the simplex method. One MPS/360 program alternates with a Fortran program to solve the optimization problem in an iterative fashion. The Fortran program is necessary to read MPS/360 solutions and to generate new data for the next iteration. Details of the software aspects are given by Mohler and Shen [34] and by Mohler and Price [41]. Detailed information of MPS/360, its linear programming package, and its controlling system, Operating System/360, are available in IBM manuals [42–44].

3.7 OPTIMAL FEEDBACK CONTROL

The optimal control problem for linear systems with the integrand of the performance index quadratic in state and control, yields a nice linear feedback solution. The feedback gain matrix is obtained as a solution of an appropriate matrix Ricatti equation [12, 20]. For bilinear systems, however, it generally is not so easy to get a closed-form solution. Again, various sequential algorithms can be developed to solve the problem. A classic method used in practice involves linearization of the system of equations about some desired trajectory. Then the classical linear problem with quadratic performance is solved. If the desired trajectory is not constant, the perturbation equations are time variant but still linear. The required feedback gain matrix is obtained as a time-variant solution to the appropriate matrix Ricatti differential equation. This method is applied to the controller design for a nuclear reactor by Mohler and Shen [34].

Leeper and Mulholland [45] provide an interesting technique to get a closed-form solution to a simple bilinear problem. Again assume for simplicity that the bilinear system is of first order with scalar control:

$$dx/dt = ax + bux + cu, \tag{3.111}$$

and that the performance index, nonquadratic in this case, includes a measure of error and control cost in the form of

$$J = \int_{t_0}^{\infty} [h(\rho) + \dot{\rho}^2] \, d\rho. \tag{3.112}$$

Here, $\rho = qx^2$, $q > 0$, $h(\rho)$ is a positive real-valued function of ρ, and $h(0) = 0$. The exact form of the performance index makes the problem solution quite convenient. In general, the method may be applied to higher order systems with ρ a positive-semidefinite form that provides a measure of system error. Hence, ρ gives an indirect measure of control energy.

The first step with this method is to measure the performance index with disregard of the system equations. Then the control is selected so that the system equations are satisfied for the extremal trajectory. The extremal trajectory must satisfy the Euler–Lagrange equation and the associated boundary conditions. For (3.112), this yields

$$\frac{d}{dt}\frac{\partial f_0}{\partial \dot{\rho}} - \frac{\partial f_0}{\partial \rho} = 0,$$

where $f_0 = h(\rho) + \dot{\rho}^2$. This reduces to

$$2\frac{d^2\rho}{dt^2} - \frac{dh(\rho)}{\partial \rho} = 0, \tag{3.113}$$

where $\rho(t_0) = qx^2(t_0)$ and $\lim_{t\to\infty} \dot{\rho}(t) = 0$. Multiplication by $\dot{\rho}$ and direct integration with the specified conditions show that

$$\dot{\rho} = -[h(\rho)]^{1/2}, \tag{3.114}$$

where the negative sign is chosen to minimize (3.112).

Then, substitution of $\rho = qx^2$ yields

$$\dot{x} = -\tfrac{1}{2}x$$

or

$$ax(t) + bx(t)u(t) + cu(t) = -\tfrac{1}{2}x(t).$$

Consequently, the desired control is

$$u = \frac{(a + \frac{1}{2})x}{bx + c}. \tag{3.115}$$

3.8 CONCLUSIONS

As a consequence of their variable structure, bilinear systems can be designed to provide relatively good performance. For example, it is sometimes possible to add a bilinear mode of control to a linear system to improve system performance by a significant amount. In this chapter, sufficiency conditions have been presented for optimal bilinear regulation. Two new computational algorithms, switching-time-variation method (STVM) and quasi-linear programming, have been introduced. The first is particularly effective for many bang–bang bilinear control problems, and the latter for problems with state constraints.

The STVM uses well-developed solutions for linear systems, and in this manner provides a relatively efficient algorithm. Also, it has been shown that solutions obtained by the STVM converge to a solution of the two-point boundary value problem associated with the maximum principle. Still there may be multiextremal solutions, and the worth of the STVM as well as quasi-linear programming sometimes depends on the ability to provide a good initial guess at the optimal trajectory. In practice, such guesses may be obtained frequently by solving a very simple representation of the problem by such analytical methods as the maximum principle.

Singular solutions for the most part have been neglected here; only an indication of how they arise has been presented. More detailed discussions of singular solutions and their role in bilinear systems are presented by Elliott and Tarn [32], Johnson [46], and Buyakas [47].

Exercises

3.1 Why is it necessary to specify optimal bilinear regulation in Section 3.2 with respect to a terminal hypersphere rather than the origin?

3.2 Explain the reason that the suboptimal bilinear synthesis which was analyzed in Section 3.2.2 performs so much better than the linear system as α increases in value.

3.3 For the prompt neutron kinetics

$$dn/dt = (u/l)n,$$

where $n(0) = n_0$, $u(0) = 0$, $l = 10^{-5}$ sec, $|u| \leq 10^{-3}$, and $|\dot{u}| \leq 10^{-4}\ \text{sec}^{-1}$, compute the time-optimal reactivity $u(t)$ to drive the process to an equilibrium terminal state with $n(t)_f = 10^6 n_0$.

3.4 With the solution obtained in Exercise 3.3 as an initial iteration trajectory, outline the procedure and write a computer program to compute the time-optimal trajectory to drive the single precursor kinetics

$$dn/dt = [(u - \beta)/l]n + \lambda c$$
$$dc/dt = (\beta/l)n - \lambda c,$$

from one equilibrium state to another. There, $\beta = 0.0065$ and $\lambda = 0.4\ \text{sec}^{-1}$.

3.5 Suppose the prompt neutron kinetics described in Exercise 3.3 are coupled to a heat exchange process with temperature T_u described by

$$dT_u/dt = n/c_m - awT_u,$$

where $c_m = 1000$, $a = 10$, $T_u(0) = 0 = \dot{T}_u(0)$, $T_u(t_f) = 5000$, $\dot{T}_u(t_f) = 0$, and coolant flow rate is constrained by $10 \leq w \leq 1000$. Compute the minimal coolant consumption control trajectories.

3.6 Discuss the advantages and disadvantages of the switching-time-variation method and quasi-linear programming for bilinear systems in comparison with other techniques commonly used to compute optimal trajectories.

3.7 Suppose the searchlight time-optimal control problem given in Section 3.3.3 only utilizes additive control [i.e., $u_2 = 0$ in (3.46)]. Compute the time-optimal control to drive the state to the origin in minimal time.

3.8 Compare the performance of the time-optimal trajectories derived for Exercise 3.7 with those in Section 3.3.3.

3.9 Explain the homogeneity property of bilinear systems without additive control and discuss its computational utility.

APPENDIX A SUFFICIENT CONDITION OF OPTIMALITY

The sufficiency theorem utilized in Section 3.2 is proven here. This proof follows Mohler and Rink [25] and is a generalization of Boltyanskii's derivation [19] for a smooth terminal $(n - 1)$-fold (hypersurface) $\theta_f \in R^n$, described by $\theta_f(\mathbf{x}) = \{\mathbf{x} \mid \mathbf{g}(\mathbf{x}) = \mathbf{0}\}$. Then, if $\{\mathbf{x}(t) \mid t \in [t_0, t_f]\}$ is any extremal trajectory terminating on θ_f, the corresponding terminal costate vector $\mathbf{p}(t_f)$ must be a scalar multiple of the unique normal to θ_f at $\mathbf{x}(t_f)$ which is $\partial\mathbf{g}/\partial\mathbf{x}(\mathbf{x}(t_f))$.

Now suppose that there is an extremal control synthesis defined on $R^n - \theta_f$. That is, for every state $\mathbf{x} \in (R^n - S)$, there is defined a control value $\mathbf{u}(\mathbf{x}) \in U$, with the following properties.

(a) $R^n - S$ is partitioned into a finite number of connnected cells by a set V, where V is the union of a finite number of hypersurface elements, and $\mathbf{u}(\mathbf{x})$ is continuously differentiable on $(R^n - \theta_f) - V$. (Practically, V is the "switching set" of the problem, where the control may undergo a jump discontinuity.)

(b) Starting at any initial state $\mathbf{x}$ in $R^n - \theta_f$, with initial control value $\mathbf{u}(\mathbf{x})$ and subsequent control policy $\mathbf{u}[\boldsymbol{\phi}(\tau)]$, the resulting trajectory $\mathbf{x}(\tau)$ is an extremal which intersects ϕ_f after an elasped time $t(x)$, and which intersects V with nonzero angle at a finite number of points.

Corresponding to this family of trajectories and control policies is a cost functional

$$J(x) = \int_{-t(\mathbf{x}_0)}^{0} f_0\{\boldsymbol{\phi}(\tau), \mathbf{u}(\boldsymbol{\phi}(\tau))\}\, d\tau,$$

which is a well-defined number for each $\mathbf{x} \in R^n$. Clearly, $J(\mathbf{x}) = 0$ on the target set θ, and since f_0 is positive definite, $J(x) < 0$ for all $\mathbf{x} \in (R^n - \theta_f)$. Assume further that

(c) $J(\mathbf{x})$ is continuous on R^n.

With these assumptions, the synthesis can be shown to be optimal. The first step is to show that the cost $J(\mathbf{x})$ is continuously differentiable everywhere in $(R^n - \theta_f) - V$, and that its gradient satisfies Bellman's equation,

$$\min_{u \in U} \left\langle f_0(\mathbf{x}, \mathbf{u}) \frac{\partial J}{\partial \mathbf{x}}, \mathbf{f}(\mathbf{x}, \mathbf{u}) \right\rangle = f_0[\mathbf{x}, \mathbf{u}(\mathbf{x})] + \left\langle \frac{\partial J}{\partial \mathbf{x}}, \mathbf{f}[\mathbf{x}, \mathbf{u}(\mathbf{x})] \right\rangle = 0 \tag{A.1}$$

everywhere in $(R^n - \theta_f) - V$. Since this result is hardly unexpected, the proof, which is rather lengthy, will be omitted. However, the rigorous proof is presented by Rink and Mohler [48].

Now, let $\mathbf{x}$ be an arbitrary point in $R^n - \theta_f$ and let $\hat{\mathbf{u}}(t)$ be any admissible control policy which transfers the system from $\mathbf{x}$ to θ_f in time $t(x)$, along with the trajectory $\{\hat{\boldsymbol{\phi}}(\tau) \mid \tau \in [-\hat{t}(x), 0]\}$. The cost of this policy is

$$\hat{J}(\mathbf{x}) = \int_{-\hat{t}(\mathbf{x})}^{0} f_0(\hat{\mathbf{x}}(\tau), \hat{\mathbf{u}}(\tau))\, d\tau.$$

Along those portions of the trajectory which lie in $(R^n - \theta_f) - V$, Bellman's equation is satisfied by the cost functional $J(\mathbf{x})$, so on these portions

$$f_0[\hat{\boldsymbol{\phi}}(\tau), \hat{u}(\tau)] \geq -\left\langle \frac{\partial J}{\partial \mathbf{x}} (\hat{\boldsymbol{\phi}}(\tau)), \mathbf{f}(\hat{\boldsymbol{\phi}}(\tau), \hat{\mathbf{u}}(\tau)) \right\rangle.$$

If Bellman's equation is satisfied everywhere on $\{\hat{\boldsymbol{\phi}}(\tau) \mid \tau \in [-\hat{t}(\mathbf{x}), 0]\}$, integration of both sides of the above inequality over $[-\hat{t}(\mathbf{x}), 0]$ yields the result $\hat{J}(x) \geq J(x)$. This implies that $J(\mathbf{x})$ is the minimum cost.

It is possible, however, that segments of the trajectory $\hat{\boldsymbol{\phi}}(\tau)$ lie in hypersurface elements of V where Bellman's equation does not hold. In this case, Lemma 2 of Boltyanskii [19, p. 334] can be invoked, as follows: According to this lemma, there exists in any ε-neighborhood of $\mathbf{x}$ a point $\mathbf{y}$ such that the trajectory originating at y with control policy $\hat{\mathbf{u}}(\tau)$ intersects V at only the isolated points in time $\tau_1, \tau_2, \ldots, \tau_r$. Let this perturbed trajectory be denoted by $\{\tilde{\boldsymbol{\phi}}(\tau) \mid \tau \in [-\hat{t}(\mathbf{x}), 0]\}$. Since the solution of the state dynamical equations depends continuously on initial conditions, and f_0 is continuous, there exists a constant k such that

$$\int_{-t(\mathbf{x})}^{0} \{f_0[\tilde{\boldsymbol{\phi}}(\tau), \hat{\mathbf{u}}(\tau)] - f_0[\hat{\boldsymbol{\phi}}(\tau), \hat{\mathbf{u}}(\tau)]\}\, d\tau \leq k\varepsilon$$

for every $\varepsilon > 0$. On the intervals $(-\hat{t}(\mathbf{x}), \tau_1)$, $(\tau_1, \tau_2), \ldots, (\tau_r, 0)$ Bellman's equation implies that

$$f_0[\tilde{\boldsymbol{\phi}}(\tau), \hat{\mathbf{u}}(\tau)] \geq \left\langle \frac{\partial J[\tilde{\boldsymbol{\phi}}(\tau)]}{\partial \mathbf{x}}, f[\tilde{\boldsymbol{\phi}}(\tau), \hat{\mathbf{u}}(\tau)] \right\rangle.$$

Also, $J(\mathbf{x})$ is continuous at the points $\boldsymbol{\phi}(\tau_1), \boldsymbol{\phi}(\tau_2), \ldots, \boldsymbol{\phi}(\tau_r)$, so

$$\begin{aligned}
&\int_{-t(\mathbf{x})}^{0} f_0[\tilde{\boldsymbol{\phi}}(\tau), \hat{\mathbf{u}}(\tau)]\, d\tau \\
&\quad \geq \left[\int_{-t(\mathbf{x})}^{\tau_1} + \int_{\tau_1}^{\tau_2} + \cdots + \int_{\tau_r}^{0}\right] \left\langle \frac{\partial J[\tilde{\boldsymbol{\phi}}(\tau)]}{\partial \mathbf{x}}, f[\tilde{\boldsymbol{\phi}}(\tau), \hat{\mathbf{u}}(\tau)] \right\rangle d\tau \\
&\quad = J(\mathbf{x}) - J[\tilde{\boldsymbol{\phi}}(0)].
\end{aligned}$$

Combining the foregoing estimates results in

$$\begin{aligned}
\hat{J}(\mathbf{x}) &= \int_{-\hat{t}(\mathbf{x})}^{0} f_0[\hat{\boldsymbol{\phi}}(\tau), \hat{\mathbf{u}}(\tau)]\, d\tau \\
&\geq \int_{-\hat{t}(\mathbf{x})}^{0} f_0[\tilde{\boldsymbol{\phi}}(\tau), \hat{\mathbf{u}}(\tau)]\, d\tau - k\varepsilon \\
&\geq J(\mathbf{x}) - J[\tilde{\boldsymbol{\phi}}(0)] - k\varepsilon
\end{aligned}$$

for every $\varepsilon > 0$. Letting $\varepsilon \to 0$, the terminal point $\tilde{\mathbf{x}}(\tilde{\boldsymbol{\phi}})$ approaches S, so $J[\tilde{\boldsymbol{\phi}}(0)] \to 0$. Therefore $\hat{J}(\mathbf{x}) \geq J(\mathbf{x})$, which implies that $J(\mathbf{x})$ is the minimum achievable cost. It is concluded that, if θ_f is a smooth hypersurface, an extremal synthesis which satisfies conditions (a), (b), and (c) is in fact optimal.

APPENDIX B OPTIMUM NUMBER OF SWITCHINGS FOR STVM

To discuss the computation of the optimum number of switchings, define the following sets: $Z^N = \{j: \hat{\tau}_j^N \neq \hat{\tau}_i^N\}$, the set containing the indices of distinct elements of $\hat{\boldsymbol{\tau}}^N$; $Z_\tau^N = \{\hat{\tau}_j^N : j \in Z^N\}$, the set containing distinct elements of $\hat{\boldsymbol{\tau}}^N$; $Z_\phi^N = \{t: \hat{\phi}^N(t) = 0, t \in (t_0, t_f)\}$, where $\hat{\boldsymbol{\tau}}_i^N$ and $\hat{\tau}_j^N$ are elements of $\hat{\tau}^N$ as defined in (3.34), N is the given number of switchings, and $\hat{\phi}^N(t)$ is the function $\phi(t) = 2\langle \boldsymbol{\lambda}(t), \mathbf{b}(x(t))\rangle$, which corresponds to $\hat{\boldsymbol{\tau}}^N$.

It follows from (3.34) that

$$\phi_j(\hat{\boldsymbol{\tau}}^N) = (-1)^j \hat{\phi}^N(\hat{\tau}_j^N) = 0 \qquad \text{for} \quad j \in Z^N, \tag{B.1}$$

where ϕ_j is the jth component of $\boldsymbol{\phi}$ [29].

If the number of elements in Z_τ^N equals R, then a new switching vector $\boldsymbol{\tau}^R = (\tau_1^R, \ldots, \tau_R^R) \in T^R$ and a new control $u^R(t)$ can be defined as $\tau_i^R = \hat{\tau}_j^N$ $(i = 1, \ldots, R), j \in Z^N$, and $u^R(t) = v(\tau^R)$ such that

$$u^R(t) = \hat{u}^N(t) \qquad \text{almost everywhere,}$$

where $\hat{u}^N(t) = v(\hat{\boldsymbol{\tau}}^N)$. Then, it can be seen that τ_i^R $(i = 1, \ldots, R)$ are all distinct, and $\hat{u}^R(t)$ differs from $u^N(t)$ only on the set consisting of the finite number of elements, $\hat{\tau}_k^N/Z^N$.

It follows that $Z_\tau^N \subset Z_\phi^N$, because, if $Z_\tau^N \not\subset Z_\phi^N$, then there must exist $t_k \in Z_\tau^N$ such that $\hat{\phi}^N(t_k) \neq 0$; that is, $\phi_k(\hat{\boldsymbol{\tau}}^N) \neq 0$, which is a contradiction. Hence suppose that Z_τ^N is strictly included in Z_ϕ^N. Then there must exist at least one $t_k \in (t_0, t_f)$ such that $\hat{\phi}^N(t_k) = 0$, but $t_k \notin Z_\tau^N$. That is, if $Z_\phi^N - Z_\tau^N \neq \varnothing$ (here $\varnothing$ is the empty set), then there exist $t_k \in (t_0, t_f)$ and j, $0 \le j \le R$, such that $\hat{\phi}^N(t_k) = 0$, $\tau^R < t_k < \tau_{j+1}^R$. Also, since $u^R(t) = -\operatorname{sgn}(\hat{\phi}^N(t))$ for $t \in (\tau_j^R - \varepsilon, \tau_j^R + \varepsilon)$, where $\varepsilon > 0$ is a sufficiently small number, the number of zeros of $\hat{\phi}^N(t)$ on the interval (τ_j^R, τ_{j+1}^R) must always be even.

Let $\hat{N} = R_N + 2r$, where $2r$ is the number of elements in $Z_\phi^N - Z_\tau^N$. Then a new N-dimensional switching vector

$$\boldsymbol{\tau}_0^{\hat{N}} = (\tau_{01}^{\hat{N}}, \ldots, \tau_{0\hat{N}}^{\hat{N}})^{\mathrm{T}} \tag{B.2}$$

can be defined by

$$\begin{aligned}
\tau_{0j}^{\hat{N}} &= \tau_j^R \qquad (j = 1, \ldots, a_1)\\
\tau_{k_1} < \tau_{0a_1+1}^{\hat{N}} &= \tau_{0a_1+2}^{\hat{N}} < \tau_{k_2}\\
\tau_{0j+3}^{\hat{N}} &= \tau_j^R \qquad (j = a_1 + 1, \ldots, a_2)\\
\tau_{k_3} < \tau_{0a_2+1}^{\hat{N}} &= \tau_{0a_2+2}^{\hat{N}} < \tau_{k_4}\\
\tau_{0j+2r}^{\hat{N}} &= \tau_j^R \qquad (j = a_r + 1, \ldots, R)
\end{aligned} \tag{B.3}$$

where a_i $(i = 1, \ldots, r)$ are integers such that $0 \le a_i \le \cdots \le a_r \le R$.

Then, it can be seen from (B.2) and (B.3) that $\boldsymbol{\tau}_0^{\hat{N}} \in T^{\hat{N}}$, $u^{\hat{N}}(t) =$

$u^R(t)$ almost everywhere, where $u^{\hat{N}}(t) = v(\boldsymbol{\tau}_0{}^{\hat{N}})$, and $I(\boldsymbol{\tau}_0{}^{\hat{N}}) = I(\boldsymbol{\tau}^R) = I(\hat{\boldsymbol{\tau}}^N)$. Also, it can be seen that $\phi_0{}^{\hat{N}}(t) = \phi^N(t)$, and $\phi_0{}^{\hat{N}}(\tau_j{}^{\hat{N}}) \neq 0$ for $\tau_j{}^{\hat{N}} \notin Z_\phi{}^N$; that is, $\boldsymbol{\phi}(\boldsymbol{\tau}_0{}^{\hat{N}}) \neq 0$.

Hence, since $\boldsymbol{\phi}(\tau_0{}^{\hat{N}}) \neq 0$, an $(\hat{N} \times \hat{N})$-dimensional nonnegative diagonal matrix K_0 can be found such that $\delta\boldsymbol{\tau}_0{}^{\hat{N}} = K_0\boldsymbol{\phi}(\tau_0{}^{\hat{N}}) \neq 0$, and

$$I(\tau_0{}^{\hat{N}} + \delta\boldsymbol{\tau}_0{}^{\hat{N}}) \leq I(\boldsymbol{\tau}_0{}^{\hat{N}}) = I(\hat{\boldsymbol{\tau}}^N). \tag{B.4}$$

The computation of a nonincreasing convergent sequence $\{I_i{}^{\hat{N}}\}$ and the corresponding sequence of switching vectors $\{\boldsymbol{\tau}_i{}^{\hat{N}}\}$ for a given number of switchings $\hat{N}$ is discussed in detail in the text. Hence, starting from $I_0{}^{\hat{N}} = I(\boldsymbol{\tau}_0{}^{\hat{N}})$ and $\boldsymbol{\tau}_0{}^{\hat{N}}$, the convergent sequences $\{I_i{}^{\hat{N}}\}$ and $\{\tau_i{}^{\hat{N}}\}$ can be computed using the gradient method.

Then, if the steps in this appendix and in the text are repeated, an integer $\hat{R}$ (corresponding to the optimum number of switchings) can be obtained such that

$$Z_\tau{}^{\hat{R}} = Z_\phi{}^{\hat{R}}. \tag{B.5}$$

That is, an integer $\hat{R}$ can be obtained such that the zeros of $\theta^{\hat{R}}(t)$ coincide with the switching times of $\hat{\boldsymbol{\tau}}^{\hat{R}}$.

APPENDIX C GRADIENT COMPUTATION

For the computation of the gradient, the variational form of the state equation (3.21) is given by

$$\delta\dot{\tilde{\mathbf{x}}}_i = \frac{\partial}{\partial\tilde{\mathbf{x}}_i}[\tilde{\mathbf{a}}(\tilde{\mathbf{x}}_i) + v_i\tilde{\mathbf{b}}(\tilde{\mathbf{x}}_i)]\delta\tilde{\mathbf{x}}_i + \tilde{\mathbf{b}}(\tilde{\mathbf{x}}_i)\,\delta v_i, \tag{C.1}$$

where $\delta\tilde{\mathbf{x}}_i(t_0) = 0$, $\tilde{\mathbf{x}}_i$ is a given state vector, $v_i = v(\boldsymbol{\tau}_i{}^N)$ is a given control, $\delta\tilde{\mathbf{x}}_i$ is the variation of $\mathbf{x}_i$, and δv_i is the variation of v_i. An adjoint vector $\boldsymbol{\lambda}_i$ is defined by

$$\dot{\tilde{\boldsymbol{\lambda}}}_i = -\partial[\tilde{\mathbf{a}}(\tilde{\mathbf{x}}_i) + v_i\tilde{\mathbf{b}}(\tilde{\mathbf{x}}_i)]^{\mathrm{T}}\boldsymbol{\lambda}_i/\partial\tilde{\mathbf{x}} \tag{C.2}$$

and

$$\tilde{\boldsymbol{\lambda}}_i(t_f) = \partial I/\partial\tilde{\mathbf{x}}|_{\tilde{\mathbf{x}}_i(t_f)}.$$

From (C.1) and (C.2) with $\delta\tilde{\mathbf{x}}_i(t_0) = 0$, $\tilde{\boldsymbol{\lambda}}_i(t_f) = \partial I/\partial\tilde{\mathbf{x}}|_{\tilde{\mathbf{x}}_i(t_f)}$ it follows that

$$\Delta I_i = \langle \partial I/\partial\tilde{\mathbf{x}}|_{\tilde{\mathbf{x}}_i(t_f)}, \delta\tilde{\mathbf{x}}_i(t_f)\rangle = \int_{t_0}^{t_f} [\langle\tilde{\boldsymbol{\lambda}}_i, \tilde{\mathbf{b}}(\tilde{\mathbf{x}}_i)\rangle\delta v_i]\, dt. \tag{C.3}$$

The variation of the control, δv_i, can be obtained from (3.25) with t_0 and t_f fixed or $\delta\tau_{i0} = \delta\tau_{iN+1}^N = 0$. Hence,

$$\delta v_i = 2\sum_{j=1}^{N} (-1)^{j+1}\delta(t - \tau_{ij+1}^N)\delta\tau_{ij+1}^N, \tag{C.4}$$

where $\delta(t - \tau)$ is the Dirac delta function occurring at $t = \tau$, and $\delta\tau_{ij}^N$ $(j = 0, \ldots, N + 1)$ are the variations of the switching times τ_{ij}^N. If (C.4) is substituted into (C.3), it is seen that

$$\Delta I_i = \sum_{j=1}^{N} (-1)^{j+1}\delta\tau_{ij}^N\phi_{ij}^N = -\langle\boldsymbol{\phi}_i^N, \delta\boldsymbol{\tau}_i^N\rangle \tag{C.5}$$

where

$$\phi_{ij}^N = \phi_i(\tau_{ij}^N), \qquad j = 1, \ldots, N$$

$$\phi_i(t) = 2\langle\tilde{\boldsymbol{\lambda}}_i(t), \tilde{\mathbf{b}}(\tilde{\mathbf{x}}_i(t))\rangle$$

$$\boldsymbol{\phi}_i^N = (-\phi_{i1}^N, \phi_{i2}^N, \ldots, (-1)^N\phi_{iN}^N)^{\mathrm{T}}$$

and

$$\delta\boldsymbol{\tau}_i^N = (\delta\tau_{i1}^N, \ldots, \delta\tau_{iN}^N)^{\mathrm{T}}.$$

Also, since $\delta\tilde{\mathbf{x}}(t_f) = (\partial\tilde{\mathbf{x}}_f/\partial\boldsymbol{\tau}^N)\delta\boldsymbol{\tau}^N$, where $\tilde{\mathbf{x}}_f = \tilde{\mathbf{x}}(t_f)$,

$$\Delta I_i = \langle\partial I/\partial\tilde{\mathbf{x}}, \delta\tilde{\mathbf{x}}\rangle|_{t_f} = \langle\partial I/\partial\boldsymbol{\tau}^N, \delta\boldsymbol{\tau}^N\rangle. \tag{C.6}$$

Then it follows from (C.5) and (C.6) that

$$\operatorname{grad} I(\boldsymbol{\tau}_i^N) = \partial I/\partial\boldsymbol{\tau}^N(\boldsymbol{\tau}_i^N) = -\boldsymbol{\phi}_i^N. \tag{C.7}$$

APPENDIX D NUMBER OF SWITCHINGS FOR FUEL-OPTIMAL POLICY

Generally, the minimum cost obtained in that computation of the optimum number of switchings for use with the STVM analyzed in

Section 3.4 is discussed here. First define the following from terms used in Section 3.4:

$$Z = \{j\colon \hat{\tau}_j \neq \hat{\tau}_i \text{ for } i \neq j\}, \tag{D.1}$$
$$Z_\tau = \{\hat{\tau}_j\colon j \in Z\}, \tag{D.2}$$

$$\begin{aligned} \hat{u}^+(t) &= v^+(\hat{\boldsymbol{\tau}}), \\ \hat{u}^-(t) &= v^-(\hat{\boldsymbol{\tau}}), \\ \hat{u}(t) &= \hat{u}^+ + \hat{u}^- = v^+(\hat{\boldsymbol{\tau}}) + v^-(\hat{\boldsymbol{\tau}}) = v(\hat{\boldsymbol{\tau}}), \end{aligned} \tag{D.3}$$

$$\begin{aligned} Z_\tau^+ &= \{\text{distinct switching times of } u^+(t)\}, \\ Z_\tau^- &= \{\text{distinct switching times of } u^-(t)\}, \end{aligned} \tag{D.4}$$

$$\begin{aligned} Z_\beta^+ &= \{t\colon \beta(t) + 1 = 0\}, \\ Z_\beta^- &= \{t\colon \beta(t) - 1 = 0\}, \end{aligned} \tag{D.5}$$

and

$$Z_\beta = Z_\beta^+ + Z_\beta^-.$$

From Eqs. (3.79) and (D.2),

$$\phi_j(\hat{\boldsymbol{\tau}}) = 0 \qquad \text{for} \quad j \in Z,$$

because, if $\phi_j(\hat{\boldsymbol{\tau}}) \neq 0$, then a $k_j \neq 0$ can be picked such that $\delta\hat{\tau}_j \neq 0$, which contradicts (3.79). If $j \notin Z$, however, ϕ_j does not necessarily vanish, but in this case k_j must be zero such that (3.79) is satisfied.

It can be seen from (D.4) and (D.5) that

$$Z_\tau^+ \subseteq Z_\beta^+ \qquad (\text{or} \quad Z_\tau^- \subseteq Z_\beta^-), \tag{D.6}$$

for if $Z_\tau^+ \not\subseteq Z_\beta^+$, then there exists some $\hat{\tau}_a \in Z_\tau^+$ such that $\beta(\hat{\tau}_a) + 1 \neq 0$, which implies that $\phi_a(\hat{\boldsymbol{\tau}}) \neq 0$ and $\delta\hat{\tau}_a \neq 0$. But this is a contradiction.

Suppose Z_τ^+ is strictly included in Z_β^+ (or $Z_\tau^- \subset Z_\beta^-$). Then there exists a $t_a \in [t_0, t_f]$ such that $t_a \notin Z_\tau^+$ (or $t_a \notin Z_\tau^-$), but $\beta(t_a) + 1 = 0$ [or $\beta(t_a) - 1 = 0$]. Since $t_z \notin Z_\tau^+$ (or $t_a \notin Z_\tau^-$), there exists j ($1 \leq j \leq N^+$) such that $\beta(t_a) + 1 = 0$ and

$$\hat{\tau}_{j-1} < t_a < \hat{\tau}_j, \tag{D.7}$$

where $\hat{\tau}_{j-1}$ and $\hat{\tau}_j$ are elements of Z_τ^+. Also since it can be shown that

$$\hat{u}^+(t) = \tfrac{1}{2}\{-\text{sgn}[\beta(t) + 1] + 1\}, \qquad t \in (\tau_k - \varepsilon, \tau_k + \varepsilon), \quad \text{for} \quad k = j - 1, j, \tag{D.8}$$

where $\varepsilon > 0$ is a sufficiently small number, the number of zeros of $\beta(t) + 1$ on $[\hat{\tau}_{j-1}, \hat{\tau}_j]$ must always be even. Hence assume that there are $2R^+$ $(2R^-)$ such zeros of $\beta(t) + 1$ [or $\beta(t) - 1$] on $[t_0, t_f]$.

Then, define a new switching vector $\boldsymbol{\tau}_0$, which consists of those new $2(R^+ + R^-)$ switching times and those switching times in Z such that

$$\boldsymbol{\tau}_0 \in T^{\hat{N}+R}$$

and

$$v(\boldsymbol{\tau}_0) = \hat{u}(t) \qquad \text{almost everywhere,}$$

where $R = R^+ + R^-$, and $\hat{N}$ is the number of elements in Z. Then, $\boldsymbol{\phi}(\boldsymbol{\tau}_0) \neq \mathbf{0}$ and $\delta\boldsymbol{\tau}_0 \neq \mathbf{0}$. Hence, the cost corresponding to this new switching vector $\boldsymbol{\tau}_0$ can be further decreased.

The computations of a nonincreasing convergent sequence of cost functionals and the corresponding sequence of switching vectors, for a given number of switchings, are discussed in Section 3.4.2. Starting from $\boldsymbol{\tau}_0$ and $I(\boldsymbol{\tau}_0)$, the limit of the convergent sequence of cost functionals and the corresponding switching vector can be obtained.

Then, if the steps given here and in Section 3.4 are repeated, a nonnegative integer R can be obtained such that $Z_\tau^R = Z_\beta^R$.

It is shown in Section 3.3.2 that the solution obtained by the STVM satisfies the TPBVP derived from the maximum principle for problems with cost functionals which are linear in control. A similar proof can be carried out for fuel-optimal control problems. The only difference is that the control for a fuel-optimal control problem is given by

$$u^0(t) = \text{dez}[\alpha^0(t)].$$

Hence, it must be proved, for a fuel-optimal control problem, that

$$u^0(t) = \hat{u}(t), \tag{D.9}$$

where $\hat{u}(t) = v(\hat{\boldsymbol{\tau}}^R)$, the control corresponding to $\hat{I}^R$.

Since $Z_\tau^R = Z_\beta^R$, it can be shown that [29]

$$\begin{aligned} \hat{u}^+(t) &= \tfrac{1}{2}[\text{sgn}(-\hat{\beta}(t) - 1) + 1] \\ \hat{u}^-(t) &= \tfrac{1}{2}[\text{sgn}(-\hat{\beta}(t) + 1) - 1], \end{aligned} \tag{D.10}$$

and

$$\begin{aligned} \hat{u}(t) &= \hat{u}^+(t) + \hat{u}^-(t) \\ &= \tfrac{1}{2}[\text{sgn}(-\hat{\beta} - 1) + \text{sgn}(-\hat{\beta} + 1)], \end{aligned} \tag{D.11}$$

for all $t \in [t_0, t_f]$, where $\hat{\beta}(t)$ is the function $\hat{\beta}(t)$ corresponding to $\hat{\boldsymbol{\tau}}^R$.
Then, it follows from (D.10) and (D.11) that

$$\hat{u}(t) = \operatorname{dez}(-\hat{\beta}(t)). \tag{D.12}$$

But, since

$$\hat{\beta}(t) = \left\langle \boldsymbol{\lambda}, \frac{\mathbf{b}}{e_1} \right\rangle = \frac{-\langle \mathbf{p}, \mathbf{b} \rangle}{e_1 |p_0|} = -\alpha^0(t),$$

where p_0 is assumed negative, (D.12) can be replaced by

$$\hat{u}(t) = \operatorname{dez}(\alpha^0(t)) = u^0(t).$$

REFERENCES

1. Bolza, O., "Lectures on the Calculus of Variations." Dover, New York, 1961.
2. Gel'fand, I. M., and Fomin, S. V., "Calculus of Variations." Prentice-Hall, Englewood Cliffs, New Jersey, 1963.
3. Hestenes, M. R., "Calculus of Variations and Optimal Control Theory." Wiley, New York, 1966.
4. Merriam, III, C. W., "Optimization Theory and the Design of Feedback Control Systems." McGraw-Hill, New York, 1964.
5. Leitmann, G., ed., "Optimization Techniques." Academic Press, New York, 1962.
6. Bellman, R. E., "Dynamic Programming." Princeton Univ. Press, Princeton, New Jersey, 1957.
7. Pontryagin, L. S., Boltyanskii, V. G., Gamkrelidze, R. V., and Mischenko, E. F., "The Mathematical Theory of Optimal Processes." Wiley (Interscience), New York, 1962.
8. Bellman, R. E., and Kalaba, R. E., "Quasilinearization and Nonlinear Boundary Value Problems." Amer. Elsevier, New York, 1965.
9. Rosen, J. B., The gradient projection method for nonlinear programming, I. Linear constraints. *SIAM J. Appl. Math.* **8**, 181–217 (1960).
10. Fletcher, R., and Powell, M. J. D., A rapidly convergent descent method for minimization. *Comput. J.* **6**, 163–168 (1963).
11. Gass, S. I., "Linear Programming: Methods and Applications." 2nd ed. McGraw-Hill, New York, 1964.
12. Dantzig, G. B., "Linear Programming and Extensions." Princeton Univ. Press, Princeton, New Jersey, 1963.
13. Zoutendijk, G., Nonlinear programming: A numerical survey. *SIAM J. Control* **4**, 194–210 (1966).
14. Canon, M. D., Cullum, Jr., C. D., and Polak, E., "Theory of Optimal Control and Mathematical Programming." McGraw-Hill, New York, 1970.

15. Kalman, R. E., The theory of optimal control and the calculus of variations. RIAS Rep. 61–3. Research Inst. for Advanced Studies, Baltimore, Maryland, 1961.
16. Dreyfus, S. E., "Dynamic Programming and the Calculus of Variations." Academic Press, New York, 1965.
17. Lee, E. B., and Markus, L., "Foundations of Optimal Control Theory." Wiley, New York, 1967.
18. Cesari, L., Existence theorems for optimal solutions of Pontryagin and Lagrange problems. *SIAM J. Control* **3**, 475–498 (1965).
19. Boltyanskii, V. G., Sufficient conditions for optimality and the justification of the dynamic programming method. *SIAM J. Control* **4**, 326–361 (1966).
20. Athans, M., and Falb, P. L., "Optimal Control." McGraw-Hill, New York, 1966.
21. Hermes, M., and Haynes, G., On the nonlinear control problem with control appearing linearly. *SIAM J. Control* **1**, 85–108 (1963).
22. Kelley, H. J., A transformation approach to singular subarcs in optimal trajectory and control problems. *SIAM J. Control* **2**, 234–240 (1965).
23. Hermes, H., Controllability and the singular problem. *SIAM J. Control* **2**, 241–260 (1965).
24. Balakrishnan, A. V., On the controllability of nonlinear systems. *Proc. Nat. Acad. Sci. U.S.A.* **55**, 465–468 (1966).
25. Mohler, R. R., and Rink, R. E., Control with a multiplicative mode. *J. Basic Engrg.* **91**, 201–206 (1969).
26. Mohler, R. R., and Rink, R. E., Multivariable bilinear system control. Control Systems, Vol. 2" (C. T. Leondes, ed.). Academic Press, New York, 1966.
class of nonlinear control processes. *Proc. IFAC Symp. Syst. Engrg. Approach to Comput. Control, Kyoto, 1970.*
28. Russel, D. L., Penalty functions and bounded phase coordinate control. *SIAM J. Control* **2**, 409–422 (1964).
29. Moon, S. F., and Mohler, R. R., Optimal control of bilinear systems and systems linear in control. Rep. No. EE-164 (69) NSF-118. Univ. of New Mexico, Albuquerque, 1969.
30. Cadzow, J. A., and Martens, H. R., "Discrete-Time and Computer Control Systems." Prentice-Hall, Englewood Cliffs, New Jersey, 1970.
31. Halkin, H., A maximum principle of the Pontryagin type for systems described by nonlinear difference equations. *SIAM J. Control* **4**, 90–111 (1966).
32. Elliott, D. L., and Tarn, T. J., Controllability and observability for bilinear systems. *Proc. SIAM Nat. Meeting, Univ. of Washington, Seattle, 1971.*
33. Bellman, R. E., and Kalaba, R. E., "Quasilinearization and Nonlinear Boundary Value Problems." Amer. Elsevier, New York, 1965.
34. Mohler, R. R., and Shen, C. N., "Optimal Control of Nuclear Reactors." Academic Press, New York, 1970.
35. Mohler, R. R., Moon, S. F., and Price, H. J., Optimal control computations for nuclear reactors. *In* "Computing Methods in Optimization Problems,

II" (L. A. Zadeh, L. W. Nustadt, and A. V. Balakrishnan, eds.). Academic Press, New York, 1969.
36. Gass, S. I., "Linear Programming: Methods and Applications," 2nd ed. McGraw-Hill, New York, 1964.
37. Dantzig, G. B., "Linear Programming and Extensions." Princeton Univ. Press, Princeton, New Jersey, 1963.
38. Hadley, G., "Linear Programming." Addison-Wesley, Reading, Massachusetts, 1962.
39. Zadeh, L. A., and Whalen, B. H., On optimal control and linear programming. *IRE Trans. Automat. Control* **AC-7**, 45–46 (1963).
40. Rosen, J. B., Iterative solution of optimal control problems. *SIAM J. Control* **4**, 233–244 (1966).
41. Mohler, R. R., and Price, H. J., Optimal reactor control. Rep. No. EE-142 (67). Univ. of New Mexico, Albuquerque, 1967.
42. Mathematical programming system/360 (369A-CO-14x) control language user's manual. IBM Rep. IBM, New York, 1966.
43. Mathematical programming system/360 (360A-CO-14x) linear programming user's manual. IBM Rep. IBM, New York, 1966.
44. IBM system/360 operating system job control language. IBM Rep. IBM, New York, 1966.
45. Leeper, J. L., and Mulholland, R. J., Optimal control of nonlinear single input systems. *IEEE Trans. Automat. Control* **AC-17**, 40 (1972).
46. Johnson, C. D., Singular solutions in optimal control. *In* "Advances in Control Systems, Vol. 2 " (C. T. Leondes, ed.). Academic Press, New York, 1966.
47. Buyakas, V. I., Optimal control of systems with variable structure. *Automat. Remote Control* **27**, 579–589 (1966).
48. Rink, R. E., and Mohler, R. R., "Controllability and Optimal Control of Bilinear Systems." Prentice-Hall, Englewood Cliffs, New Jersey, 1970.

IV

Nuclear and Thermal Control Processes

Diminishing sources of conventional fuel, coupled to rapidly growing demands for power, are once again bringing nuclear power to the forefront. The significant role of nuclear fission and, eventually, fusion is emphasized even further by an ecologically conscious society. Consideration of cost–effectiveness and pollution points at the need for high performance or optimum design and operation.

Since it is doubtful that a controlled fusion process will be a competitive power source during the next two decades, consideration here is given only to control of neutron-induced fissions of isotopes of uranium, plutonium, or thorium. The fission process, carefully regulated in the form of a nuclear chain reactor, produces heat as a useful output. Heat exchangers, which are required to cool the reactor and to provide the thermal energy to any required converter, may be quite complex in design. The basic reactor may have a solid, liquid, or gaseous core with direct or indirect heat transfer arranged by convection, conduction, and/or radiation. Probably the most common design involves a solid core with direct transfer of heat to a moving fluid.

Within the reactor itself, fissions of the fuel nucleus take place mainly in the core. The core normally is surrounded by two concentrically surrounding shells which form a neutron reflector and a radiation-absorbing shield. The fission process usually is made more efficient by adding a moderator to the core so that the energy of the fission neutrons is decreased by inelastic scatterings. This in turn increases the probability of neutrons fissioning fuel nuclei. For some applications, however, the added weight and design complexity due to the moderation are not justified and the configuration is called a fast reactor. Fertile material such as ^{238}U can be added to certain fast-fission reactors to breed more fuel and yet augment reactor power by as much as 25 percent. Such reactors that produce more fissionable material than they deplete are termed breeder systems. Unfortunately, breeder reactors are quite expensive and require advanced material technology.

An electric power generation station may consist of a nuclear reactor (eventually a fast-fission breeder type), a direct heat exchanger which transfers fission heat to the coolant of a primary cycle, and an indirect heat exchanger to transfer coolant heat to cycling water. The latter, in turn, may drive a steam turbine.

For space stations, nuclear power sources are ideal because of their long lifetime and high power density. Indeed, space stations must be reliable and lightweight. Consequently, nuclear power stations may use thermal-electric energy converters such as thermionic units, thermoelectric devices, fuel cells, or magnetohydrodynamic (MHD) systems. Any optimum allocation of these converters to a plant would depend on overall system requirements. Magnetohydrodynamic systems [1] and fuel cells may be quite efficient, but they are also generally heavy. This usually leads to a compromised design according to overall specifications. An ultrahigh heat transport device, called a heat pipe [2], seems to be most amenable to electric power generation in space [3].

Nuclear propelled rockets with their large specific impulses (thrust per unit of coolant mass flow rate) will be very efficient for interplanetary space missions [4]. For example, with present technology nuclear rockets can obtain specific impulses of about 1000 sec compared to about 400 sec at the most for chemical systems.

Bilinear models for nuclear fission and for heat transfer are studied in this chapter. Then optimal neutronic control, optimal heat transfer control, and control of nuclear power systems are analyzed. A classic linear feedback design is used to approximate an optimal configuration.

4.1 NUCLEAR FISSION

When a fertile nucleus undergoes fission, an average of two or three neutrons are emitted along with nuclear radiation and a relatively large amount of energy. The energy causes rapid motion of fission fragments which produce heat. Another, though much smaller, amount of heat is generated as a consequence of the decay of fission products. As a result, neutron flux is nearly proportional to heat power except during rapid shutdowns. For sudden large flux decreases, heat power may drop rapidly to a few percent of operating level due to fission heat, but the remaining heat decays more slowly, at a rate determined by the decay of the radioactive fission products.

Generally, it is assumed that through careful core design, neutron time variations are independent of spatial variations, and that neutron energy spectrum is independent of neutron level [5]. Then the core is assumed to be a lumped source of neutrons with prompt heat power, neutron population, and neutron flux all related at a given spatial point by constants of proportionality.

A rigorous derivation of the neutronic state equations based on the time-dependent monoenergetic transport equation is given by Keepin [5, pp. 161–166]. The heuristic derivation given here follows that of Mohler and Shen [6, pp. 3–6].

The net change in neutron population over one generation is

$$dn/dt = (k - 1)n/l, \tag{4.1}$$

where k, the average number of first-generation offspring per neutron death, is called the multiplication constant; l is the mean prompt neutron generation time; and all neutrons are produced in a time $t \ll l$. Here, l may be a millisecond for thermal reactors, or might be a microsecond for fast reactors. This model assumes that all neutrons are produced promptly. It is common knowledge, however, that a small portion of neutrons are derived from unstable fission products called precursors. Six distinct groups of delayed neutrons have been observed, and their characteristics are observed in Table 4.1 for ^{235}U fuel.

Equation (4.1) for prompt neutrons may be modified to account for delayed neutrons by merely subtracting $k\beta n/l$, and for a neutron source

TABLE 4.1

^{235}U DELAYED NEUTRON PARAMETERS FOR FAST FISSION[a]

Group i	Half-life (sec)	Decay constant λ_i (sec^{-1})	Portion[b] β_i/β
1	54.51	0.0127	0.038
2	21.84	0.0317	0.213
3	6.00	0.115	0.188
4	2.23	0.311	0.407
5	0.496	1.40	0.128
6	0.179	3.87	0.026

[a] Keepin [5].
[b] $\beta = \sum_{i=1}^{6} \beta_i \approx 0.0065$.

rate s by adding s. Then the additional delayed neutrons emitted by the six precursors cause the rate of neutron change to be

$$\frac{dn}{dt} = \left[\frac{k(1-\beta)-1}{l}\right]n + \sum_{i=1}^{6} \lambda_i c_i + s, \tag{4.2}$$

where λ_i is the decay constant for the ith group of precursors, and c_i is the population of the ith precursor group. It is assumed that delayed neutrons have the same effect on the process as do prompt fission neutrons.

The rate of precursor population change equals birth rate minus death rate:

$$\dot{c}_i = dc_i/dt = (k\beta_i/l)n - \lambda_i c_i, \qquad i = 1, \ldots, 6, \tag{4.3}$$

where β_i is the portion of neutrons generated from the ith precursor. Recall that Eqs. (4.2) and (4.3) are presented as (1.4) and (1.5) with multiplicative control, $u(t) = k(t)$, and without additive control source s which is usually a relatively small constant.

The total neutron population is at constant value n_1 if

$$k = \left(l \sum_{i=1}^{6} \dot{c}_i/n_1\right) + 1,$$

and $s = 0$. At this so-called delayed critical condition the neutron kinetics, Eqs. (4.2) and (4.3), are at equilibrium state, $n(t) = n_1$ and $c_i(t) = c_{1i}$ $(i = 1, \ldots, 6)$, if $k = 1$ and $c_{1i} = (\beta/l\lambda_i)n_1$.

Ordinarily, around design level the system is operated near delayed critical with k approximately 1 and s negligible. Then the kinetics are approximated by

$$\frac{dn}{dt} = \frac{\delta k - \beta}{l} n + \sum_{i=1}^{6} \lambda_i c_i \tag{4.4}$$

and

$$dc_i/dt = (\beta_i/l)n - \lambda_i c_i, \qquad i = 1, \ldots, 6, \tag{4.5}$$

where the reactivity δk is $k - 1$.

In all of these cases the system is bilinear, and even the distributed dynamics can be described by a bilinear partial differential equation.

Frequently, the neutron kinetics are accurately approximated by a single-precursor model of the form

$$\frac{dn}{dt} = \frac{\delta k - \beta}{l} n + \lambda c \tag{4.6}$$

and

$$\frac{dc}{dt} = \frac{\beta}{l} n - \lambda c, \tag{4.7}$$

where λ is an average decay constant for an average precursor of population c.

Control of this fission process is effected by means of reactivity δk or multiplication k. Either one may be utilized as a control variable, depending on which equation is used. If $u(t) = \delta k(t)$ is constant, the eigenvalues for the system $\mathbf{A}$ matrix,

$$\mathbf{A} = \begin{bmatrix} (u - \beta)/l & \lambda \\ \beta/l & -\lambda \end{bmatrix}, \tag{4.8}$$

are

$$\rho_1, \rho_2 = \frac{(u - \beta - \lambda l) \pm [(u - \beta - \lambda l)^2 + 4u\lambda l]^{1/2}}{2l}, \tag{4.9}$$

and the corresponding eigenvectors have directions θ_1, θ_2 such that

$$\tan \theta_i = [l\rho_i - (u - \beta)]/\lambda l, \qquad i = 1, 2. \tag{4.10}$$

4.1.1 Time-Optimal Neutronic Control

The minimum-time reactivity to control the neutron fission process from some initial equilibrium state to some terminal state is computed in this section. For simplicity let the process be approximated by (4.6) and (4.7) or by

$$d\mathbf{x}/dt = \mathbf{A}(u)\mathbf{x} = \mathbf{f}(\mathbf{x}, u), \tag{4.11}$$

where $\mathbf{A}(u)$ is defined by (4.8), $x_1 = n$, and $x_2 = c$. The first problem is to find the admissible control $|u(t)| \leq \gamma$ from the class of piecewise continuous functions which drives the system from an initial equilibrium (n_0, c_0) to a terminal equilibrium (n_f, c_f) in minimal time.

Following the maximum principle [7], the optimal control must maximize the inner product

$$\begin{aligned} \mathscr{H}(n, c, \mathbf{p}, u) &= \langle \mathbf{f}(\mathbf{x}, u), \mathbf{p} \rangle \\ &= (u/l)np_1 + (p_2 - p_1)[(\beta/l)n - \lambda c], \end{aligned} \tag{4.12}$$

where $\mathbf{p}(t)$ is a continuous nonzero **n**-vector with

$$d\mathbf{p}/dt = \mathbf{A}^{\mathrm{T}}(u)\mathbf{p}, \tag{4.13}$$

$\mathbf{p}(t_0) = \mathbf{p}_0$, and generally $\mathscr{H}$ is related to H used in Chapter III by

$$H(\tilde{\mathbf{x}}, \tilde{\mathbf{p}}, \mathbf{u}) = \mathscr{H}(\mathbf{x}, \mathbf{p}, \mathbf{u}) + p_0 f_0(\mathbf{x}, \mathbf{u}). \tag{4.14}$$

Consequently, since $n(t)$ is positive, the optimal control has the form of

$$u = \gamma \operatorname{sgn} p_1(t), \tag{4.15}$$

and the maximum value of (4.14) must be nonnegative as required by the maximum principle. Substitution of $u = \pm\gamma$ into (4.6) and (4.7), with $(\beta/l)n_0 = \lambda c_0$ for initial equilibrium, yields

$$n = \frac{[\rho_2 \mp (\gamma/l)]n_0 \exp \rho_1(t - t_0) - [\rho_1 \mp (\gamma/l)]n_0 \exp \rho_2(t - t_0)}{\rho_2 - \rho_1} \tag{4.16}$$

and

$$c = \frac{\rho_2 c_0 \exp \rho_1(t - t_0) - \rho_1 c_0 \exp \rho_2(t - t_0)}{\rho_2 - \rho_1}, \tag{4.17}$$

where ρ_1, ρ_2 are given by (4.9) with $u = \pm\gamma$.

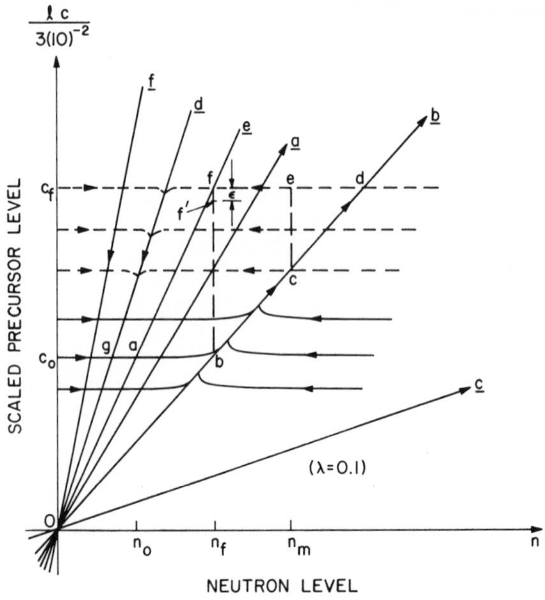

Fig. 4.1. Time-optimal neutronic state trajectories. **e**, equilibrium line ($\dot{c} = 0, \dot{n} = 0$ with $u = 0$); eigenvectors **a**, **b**, **d**, **c**, and **f** for $u = 0.25\beta$, $\pm 0.5\beta$, and $\pm 0.9\beta$, respectively [6].

Since the two eigenvalues are real and of opposite sign for $u = \gamma$, and both are negative and real for $u = -\gamma$, the bang–bang-controlled neutron kinetics are characterized in the state plane by a saddle-point portrait superimposed on a stable-node portrait with a singular point at the origin. It can be seen by elimination of time from (4.16) and (4.17) that the state trajectories are located by

$$\left[\frac{(\rho_2 + \lambda)c - (\beta/l)n}{(\rho_2 + \lambda)c_0 - (\beta/l)n_0}\right]^{-\rho_2} = \left[\frac{(\rho_1 + \lambda)c - (\beta/l)n}{(\rho_1 + \lambda)c_0 - (\beta/l)n_0}\right]^{-\rho_1}, \quad (4.18)$$

where n_0 and c_0 locate any given state. Certain such combinations of time-optimal trajectories are given in Fig. 4.1. State behavior of the adjoint system (4.13), with $u = 0.9\beta$ for positive p_1 and $u = -0.9\beta$ for negative p_1, is described by Fig. 4.2. This so-called costate in the right half of the state plane is described by a saddle-point portrait, while that in the left half-plane is typical of an unstable-node portrait.

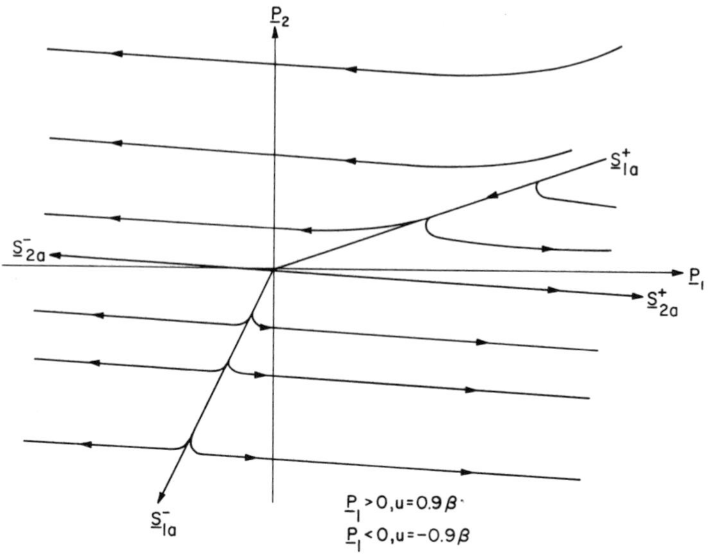

Fig. 4.2. Neutronic adjoint trajectories [6].

Since it is required that $\mathbf{p}(t)$ be continuous, it is obvious from Fig. 4.2 or from the solution of (4.13) directly that $p_1(t)$, the switching function in (4.15), cannot have more than one zero in finite time.

Since the constant-u trajectories are unique solutions of the differential equations of the system, Fig. 4.1 indicates that there is only one possible trajectory which joins any reachable state from some initial state with a maximum of one switching, and therefore such a trajectory must be optimal (e.g., trajectory a–d–f has $u = \gamma$ from a to d and $u = -\gamma$ from d to f). Also, the principle of optimality (any portion of an optimal trajectory is optimal) shows that if a desired terminal state can be reached from a given initial state with no switchings and $|u| = \gamma$, then the joining trajectory is optimal.

An example of an optimal startup trajectory linking an initial steady state (n_0, c_0) with a desired terminal steady state (n_f, c_f) is represented by trajectory a–d–f in Fig. 4.1. Initially, the system is at equilibrium with zero reactivity. Then reactivity is made equal to the maximum positive constant γ while $p_1(t)$ is positive. At point d, $p_1(t)$ goes through zero and becomes negative as reactivity is switched to the minimum constraint.

When the desired state point f is reached, reactivity is returned to zero to maintain quiescence.

The switching point for the time-optimal startup process is conveniently determined from Eq. (4.18) by the neutron level n (or the neutron precursor level c) at point d. In general, Eq. (4.18) may be used to generate a switching line for a time-optimal startup to a desired terminal state (with n_0 replaced by n_f and c_0 by c_f). Below this line, $u = \gamma$, and $u = -\gamma$ above the switching line.

If knowledge of n and c is available, an on-line computation could be made (e.g., digitally, or by means of an analog function generator and a comparator) to determine the switching point.

Since the trajectory is asymptotic to the eigenvector **b** in Fig. 4.1, for any significant difference $n_f - n_0$, the switching point is nearly independent of the initial state. In many physical systems, it might be desirable to switch to a simple control law which is a continuous function of the system state within some small predetermined region of the desired terminal state. The size and shape of this region could be determined by the accuracy of the equations in defining the physical process, the accuracy of the measuring devices, and the capability of the closed-loop control. Hence the exact switching point might not be too important.

4.1.2 Neutronic Controllability

From the bang–bang-controlled neutron kinetics and the state portrait presented by Fig. 4.1, it is convenient to analyze the problem of controllability. That is, what states can be attained in finite time with an admissible control?

Several conclusions can be reached immediately. Perhaps foremost is the fact that any state that is reachable with a piecewise continuous control, such that $|u| \leq \gamma$, is reachable with a bang–bang control of the form $u = \gamma \operatorname{sgn} F(t)$. Obviously, it is of primary interest to discuss controllability with respect to states reachable from the equilibrium line and states from which the equilibrium line is reachable.† From the discussion of the previous section (see Fig. 4.1), it is seen that every state between the dominant eigenvector for positive reactivity (see **b** in Fig.

† Here the equilibrium line refers to those states for which $dc/dt = 0$ and $dn/dt = 0$ (the latter holding if $u = 0$).

4.1) and the dominant eigenvector for negative reactivity (see **d** in Fig. 4.1) is attainable from the equilibrium set (see **e** in Fig. 4.1), and from all of these states, the equilibrium set is reachable. Naturally, the eigenvectors (including the equilibrium point at the origin) are excluded.

Consequently, it is possible to show that any state between these dominant eigenvectors (**b** and **d** on Fig. 4.1) is reachable in finite time from any other state between these eigenvectors. Since this region is controllable to and from the equilibrium line, it is only necessary to show that any equilibrium point is locally controllable. With this in mind, define the state in terms of the variation from an equilibrium point $\mathbf{x}_e$. Let $x_1 = (n - n_e)/n_e$, and $x_2 = (c - c_e)/c_e$, where c_e is an equilibrium precursor level and $n_e = l\lambda c_e/\beta$. Then Eqs. (4.6) and (4.7) become

$$l\, dx_1/dt = (u - \beta)x_1 + \beta x_2 + u \qquad \text{and} \qquad dx_2/dt = \lambda(x_1 - x_2).$$

Now, $\mathbf{x}_e^{\mathrm{T}} = (0, 0)$, and a theorem due to Markus [8] may be applied to show local controllability. Accordingly, the neutronics are locally controllable in an open neighborhood of the origin if the rank of the matrix $\mathbf{E} = [\mathbf{B}, \mathbf{AB}, \mathbf{A}^2\mathbf{B}, \ldots, \mathbf{A}^{n-1}\mathbf{B}] = n$, where

$$\mathbf{A} = \frac{\partial \mathbf{f}}{\partial \mathbf{x}}(0, 0), \qquad \mathbf{B} = \frac{\partial \mathbf{f}}{\partial \mathbf{u}}(0, 0), \qquad \text{and} \qquad \frac{d\mathbf{x}}{dt} = \mathbf{f}(\mathbf{x}, \mathbf{u}).$$

For the single-precursor neutron kinetics,

$$\mathbf{A} = \begin{bmatrix} -\beta/l & \beta/l \\ \lambda & \lambda \end{bmatrix}, \qquad \mathbf{B} = \mathbf{b} = \begin{bmatrix} 1/l \\ 0 \end{bmatrix},$$

and rank $E = n = 2$. Hence, the system is locally controllable about the equilibrium set with the exception of the origin. Therefore, with the previous argument, the system is completely controllable between the dominant positive eigenvector (**b** in Fig. 4.1) and the dominant negative eigenvector (**d** in Fig. 4.1).

Similarly, the neutron kinetics with six precursor groups are completely controllable in a seven-dimensional open region containing the equilibrium set and which has edges defined by the seven dominant eigenvectors.

By variation of a positive source term, it is possible to extend the region of controllability to include the region beyond the dominant positive eigenvectors. This may readily be shown for the single-precursor model to be the region between the eigenvector **b** and the n axis in Fig. 4.1. As shown in Chapter II, the significance of additive control with respect to controllability is very apparent.

Controllability is an important aspect of the control problem. For example, an optimal control problem is not well posed if the desired terminal state cannot be reached in finite time.

4.1.3 Time-Optimal Control of Neutron Level

Though the state control has its utility (e.g., during calibration of instrumentation when time is valuable), it is frequently desirable to traverse from some initial equilibrium state to a terminal equilibrium neutron level in minimal time without the terminal precursor level necessarily being in equilibrium. This problem may be specified by merely defining the terminal set to be the vertical line $n = n_f$ in the n vs. c state plane shown in Fig. 4.1. Here, the projection of the costate vector on state space must be perpendicular to this terminal line at the terminal time t_f.

With $p_2(t_f) = 0$ and time reversed, it is seen from Fig. 4.2 that $p_1(t)$ can have no zeros for $t < t_f$. Similar plots, with the eigenvectors in the same quadrants, could be obtained for other values of γ. Hence the time-optimal control with a reactivity constraint is constant for $t_0 < t < t_f$. That this process requires no switching on (t_0, t_f) may also be shown by the solution to the adjoint system with time reversed, which is similar to the neutronics solution.

At the terminal time, the neutron precursor level is not yet at equilibrium. Therefore, reactivity must be different from zero to maintain constant neutron level $n(t) = n_f$ for $t > t_f$. From Eq. (4.1), it is seen that

$$u = l\dot{c}/n_f = \beta - \lambda l c(t)/n_f \tag{4.19}$$

is the required terminal reactivity. From Eqs. (4.16) and (4.17) it is readily shown that this required terminal control does belong to the admissible set with $|u(t)| \leq \gamma$ [6].

If the switching function of the control equation vanishes for any finite time, the control is said to be singular. For Eq. (4.15) a singular control exists on some time interval if $p_1(t)$ is zero on this time interval. It is seen from (4.13), however, that this would require $p_2(t) = 0$.

Hence, $\mathbf{p}(t) = \mathbf{0}$ on this finite interval, which would contradict the maximum principle. Therefore, the problem does not have a singular solution. Along with uniqueness of the bang–bang-controlled state trajectories, it is obvious that the extremal neutronic policy derived here is indeed the unique time-optimal control.

If six precursor groups are considered, the state is described by Eqs. (4.4) and (4.5). The time-optimal control is still obtained from (4.15), but in this case the switching function $p_1(t)$ satisfies the adjoint equations of (4.4) and (4.5). Again, the problem of most physical interest is to control the fission process from an initial point to a terminal hyperplane, $n = n_f$, in a seven-dimensional state space. The transversality condition then requires that

$$p_i(t_f) = 0 \qquad \text{for} \quad i \neq 1.$$

From this condition it is readily seen from the homogeneity of the costate equations that $p_1(t)$ cannot have a zero in finite time. Consequently, $u(t)$ does not switch on (t_0, t_f).

Again, the terminal control for $t > t_f$ must compensate for the decaying precursors. In this case, the terminal control is

$$u = \frac{l}{n_f} \sum_{i=1}^{6} \frac{dc_i}{dt}. \tag{4.20}$$

It may be seen from the state equations that this terminal control is admissible just as for the one-precursor case.

Mohler and Shen [6] show that a small error in measurement to estimate the required terminal control results in a divergence of $n(t)$ away from n_f. A multiple bang–bang or so-called dither control† which holds neutron level essentially constant has been synthesized. The idea is to allow $n(t)$ to vary within certain small prescribed limits about n_f, and to drive the control at maximum effort about the theoretical in order to correct $n(t)$ when its deviation exceeds these limits. Figure 4.3 shows a rapid time-optimal start controlled at the terminal end by such a dither process. Here, reactor power (proportional to neutron level) is held essentially constant after a 9-msec increase by a factor of 20 in power.

† Sometimes called sliding control.

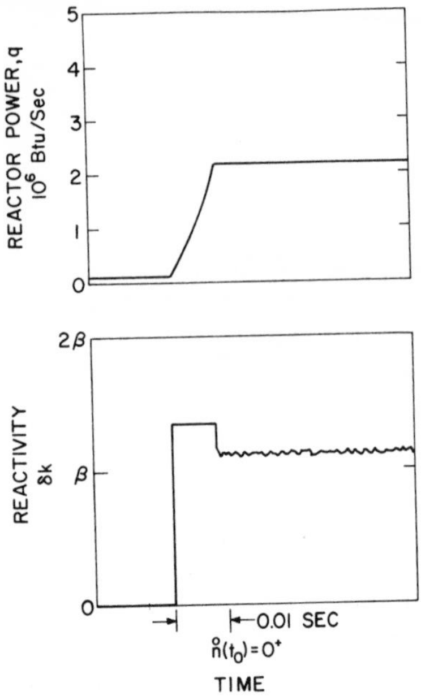

Fig. 4.3. Time-optimal neutronic transients with power constraint (see trajectory *a–c–e–f* in Fig. 4.1) [6].

4.1.4 Suboptimal Closed-Loop Control

The key to the synthesis of the time-optimal neutron level control appears to be the ability to maintain neutron density essentially constant while the precursor level is not near steady state.

Although the bang–bang neutronic control is synthesized very simply, the required terminal control is slightly more complicated. In many cases, a conventional type of closed-loop control is satisfactory and may even approach the performance of the optimal process. Even for these cases, however, the optimization analysis provides a yardstick of performance. One convenient technique of approximating the optimal process is to synthesize a continuous feedback control system which is fast-acting but limited in control variations. The startup of such a system is shown in Fig. 4.4. Initially, the system is at equilibrium, with a neutron level of less than one-twentieth of the terminal level. Then

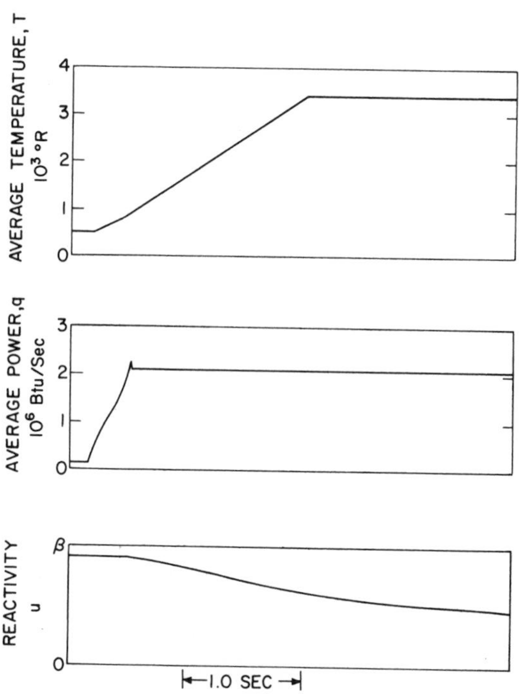

Fig. 4.4. Time-optimal neutronic start with a continuous terminal control [6].

0.9β of reactivity is added until the neutron level is about one-eighth of the terminal value. At that time, a proportional-plus-integral type of feedback controller replaces the constant control process. The controller was introduced at that time to limit the magnitude of the controller integral signal. (Log-power control or a separate integral controller limit would allow the loop to be closed for the entire startup.) The power controller used here is defined by

$$u = 1.87(10)^{-9} q_e + 6.24(10)^{-9} \int_{-\infty}^{t} q_e \, d\sigma, \tag{4.21}$$

where $q_e(t) = q_f - q(t)$ (Btu/sec) and $q \propto n$. Due to the high gain of the system, the reactivity stays at its constrained value of 0.9β until the terminal state is approached. After a small rise time, $q(t)$ has a small overshoot followed by a slow decay toward the desired steady-state value. The slow decay is caused by the slow birth of precursor neutrons. For most applications, such small overshoot is insignificant, and this simple suboptimal closed-loop control could be utilized. Nevertheless, the dither terminal control maintains constant power more accurately.

Summarizing, it is possible to design and synthesize a suboptimal neutronic control by conventional means. Knowledge of the optimal performance, however, is required to evaluate the conventional controller. Further, the small terminal transient of slow decay (which always exists for simple suboptimal control with finite gain) hardly appears if the terminal control is a closed-loop dither process.

4.1.5 Reactivity Rate Constraint

For many nuclear fission processes, it is possible to change reactivity quite rapidly, and for these, inertialess control such as that assumed in the previous sections is a good assumption. In other cases, however, control may be slow-acting due to physical inertia and energy constraints. Sometimes, reactivity is even velocity limited as well as magnitude limited for safety reasons. To apply the maximum principle (with state constraints) [7] to these problems, it is convenient to define reactivity as a new state variable, say $x_3(t)$, for a single-precursor model. Then, let the rate of change of reactivity be the control $u(t)$. Now, the neutron kinetics are described by

$$dx_1/dt = [(x_3 - \beta)/l]x_1 + \lambda x_2, \tag{4.22}$$

$$dx_2/dt = (\beta/l)x_1 - \lambda x_2, \tag{4.23}$$

and

$$dx_3/dt = u. \tag{4.24}$$

In terms of previous notation, the state variables are $x_1 = n(t)$ or $q(t)$, $x_2 = c(t)$, and $x_3 = \delta k(t)$. The dynamics are constrained so that

$$g(\mathbf{x}) = x_3^2 - \gamma^2 \leq 0 \tag{4.25}$$

and

$$|u(t)| \leq \eta. \tag{4.26}$$

For the first problem, find the admissible time-optimal reactivity variation to control the neutron kinetics from an initial equilibrium state $\mathbf{x}(t_0) = \mathbf{x}_0$ (with $x_{20} = \beta x_{10}/l\lambda$ and $x_{30} = 0$) to a desired terminal equilibrium state $\mathbf{x}(t_f) = \mathbf{x}_f$ (with $x_{2f} = x_{1f}/l\lambda$ and $x_{3f} = 0$). Again,

$$u = \eta \operatorname{sgn} p_3(t) \tag{4.27}$$

is the form of the control that makes

$$\mathscr{H}(\mathbf{x}, \mathbf{p}, \mathbf{u}) = \langle \mathbf{p}, \mathbf{f}(\mathbf{x}, \mathbf{p}, \mathbf{u}) \rangle = p_1[(x_3 - \beta/l)x_1 + \lambda x_2] + p_3 u + p_2(\beta x_1/l - \lambda x_2)$$

a maximum and is a candidate for the time-optimal strategy while $|x_3(t)| < \gamma$. The switching function is determined by $d\mathbf{p}/dt = -\partial\mathscr{H}/\partial\mathbf{x}$,

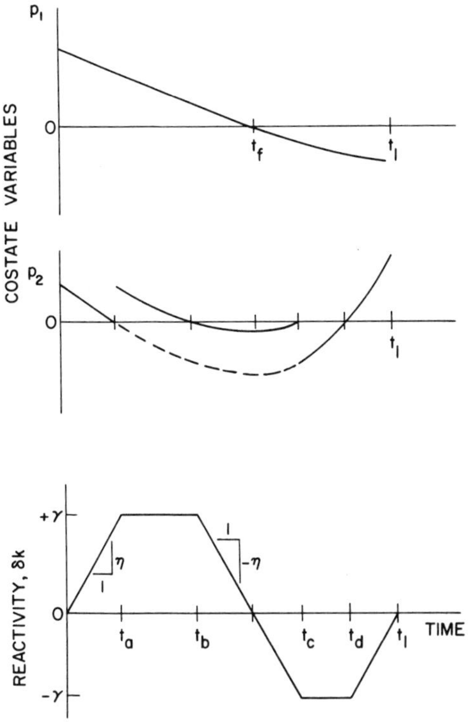

Fig. 4.5. Switching functions and maximum effort control with rate constraint [6].

or

$$dp_1/dt = (\beta/l - x_3)p_1 - (\beta/l)p_2,$$
$$dp_2/dt = \lambda(p_2 - p_1), \qquad (4.28)$$
$$dp_3/dt = -x_1 p_1.$$

For this problem, the adjoint system Eq. (4.28) describes the costate while $|x_3(t)| = \gamma$ (with $u = 0$) as well as when $|x_3(t)| < \gamma$, since (with $z = 2ux_3$) the latter part of

$$d\mathbf{p}/dt = -[\partial\mathbf{f}/\partial\mathbf{x} - \partial\mathbf{f}/\partial u(\partial\mathbf{x}/\partial u)^{-1}(\partial z/\partial\mathbf{x})^{\mathrm{T}}]^{\mathrm{T}}\mathbf{p},$$

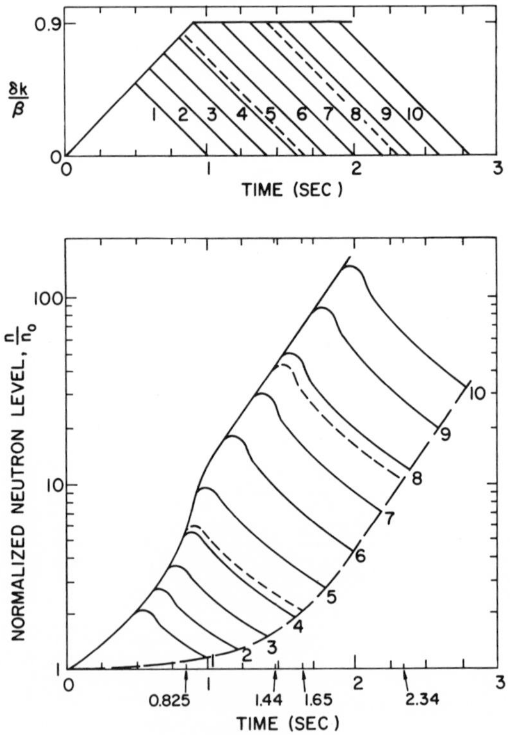

Fig. 4.6. Time-optimal neutronic control of state with reactivity rate constraint [6].

which is the addended costate equation while trajectories are on a state constraint boundary, vanishes for $u = 0$.

It is apparent that a singular control [$p_3(t) = 0$ in Eq. (4.27)] cannot satisfy the maximum principle, for, from Eq. (4.28), $p_3 = 0$ requires that $p_1 = p_2 = 0$. Thus, since the trajectories resulting from the maximum principle are unique, (4.27) is the optimal control if $|x_3| < \gamma$, and $u = 0$ while $|x_3(t)| = \gamma$.

It is shown by Mohler and Shen [6] that the switching function $p_3(t)$ can have at most four zeros as exemplified in Fig. 4.5. In practice, the negative reactivity portion is relatively smaller than shown here and usually can be neglected, since the resulting terminal error is smaller than instrumentation error. Suboptimal trajectories for several examples with the negative reactivity triangle neglected are shown in Fig. 4.6. For

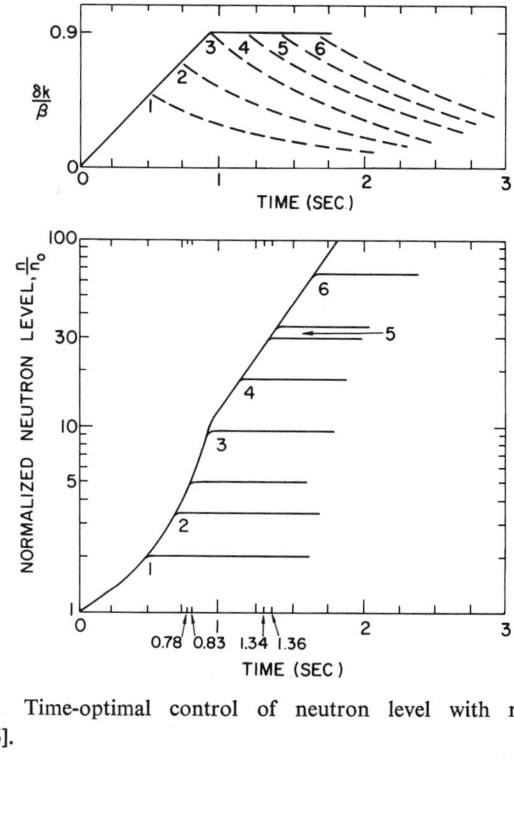

Fig. 4.7. Time-optimal control of neutron level with reactivity rate constraint [6].

these trajectories, $l = 10^{-4}$ sec, $\lambda = 0.4\ \text{sec}^{-1}$, $\gamma = 0.9\beta$, $\eta = \beta\ \text{sec}^{-1}$, and Runge–Kutta integration was utilized to obtain the data with an IBM 360 computer.

Again, it may be desired to control the kinetics from some initial equilibrium state $\mathbf{x}_0$ to some $x_1(t_f) = x_{1f}$ so that $|x_3(t)| \le \gamma$, $|u(t)| \le \eta$ for all t, and $dx_1(t)/dt = 0$ for $t \ge t_f$. From the terminal condition, $p_2(t_f) = 0$, it can be seen from (4.28) that a continuous $p_3(t)$ can have at most one zero on $[t_0, t_f]$ since $p_1(t)$ cannot have any zeros for this case (similar to the previous problem). Consequently, with a jump at the state constraint entering corner, $p_3(t)$ cannot have more than two zeros, and the time-optimal reactivity variation is a truncated trapezoid on $[t_0, t_f]$. Here t_f is much smaller than for the previous problems. Such inputs and corresponding responses are shown in Fig. 4.7. It can be seen from (4.22) to (4.24) that $x_3(t)$ is continuous at the terminal end as shown in Fig. 4.7. It is proven by Mohler and Shen [6] that adding more precursors to the model does not alter the maximum number of switchings of the control.

4.2 HEAT TRANSFER†

Heat exchanges are naturally distributed processes described by partial differential equations. In practice, however, it is common to approximate the system to some desired accuracy by finite difference models or so-called compartmented models (the latter term is used particularly in the physiological literature). Furthermore, solid reactor cores with direct heat transfer to a moving fluid such as those considered here, are carefully designed so that radial flux and temperature distributions are quite uniform. An average radial element, consequently, is utilized to approximate the process. A qualitative description of the heat transfer is provided by a single axial lump as well.

Assume that the nuclear reactor with multiple coolant flow passages is modelled by a single average cylinder. If axial heat transfer is neglected, due to small temperature gradients in the direction of coolant velocity, a thermal energy balance yields: fission heat, q (proportional

† The reader is referred to any of the many excellent heat transfer books, such as that by Eckert and Drake [9], for a more detailed discussion.

to n), equals heat stored in fuel and moderator plus heat transferred to coolant. In mathematical terms,

$$q = c_{\mathrm{m}}\, dT_u/dt + hA(T_u - T_{\mathrm{g}}), \tag{4.29}$$

where T_u is average core temperature, T_{g} is average coolant temperature, c_{m} is effective heat capacity of core mass, h is an effective heat transfer coefficient, and A is an effective heat transfer area. An energy balance on the coolant yields

$$hA(T_u - T_{\mathrm{g}}) = c_{\mathrm{mg}}\, dT_{\mathrm{g}}/dt + c_{\mathrm{g}}w(T_{\mathrm{e}} - T_{\mathrm{i}}), \tag{4.30}$$

where w is coolant mass flow rate, c_{g} is coolant mass heat capacity, T_{e} is coolant exit temperature, T_{i} is coolant entrance temperature, c_{mg} is coolant mass heat capacity, and c_{g} is coolant specific heat. The average coolant temperature is a weighted average of inlet and outlet temperatures,

$$T_{\mathrm{g}} = (T_{\mathrm{i}} + \theta T_{\mathrm{e}})/(1 + \theta), \tag{4.31}$$

where θ is a weighting constant (sometimes equal to 1 or 2).

Generated from convection, conduction, and radiation, heat transfer coefficients are generally nonlinear functions of appropriate temperatures. Of course, the effective heat transfer areas are geometrically dependent. For a solid reactor core with a fluid coolant, radiation is quite small and conduction impedance is frequently neglected compared to convective impedance. If conduction impedance is considered, usually it is assumed to be a constant dependent on thermal conduction of the material and effective thickness. The convective coefficient is strongly dependent on coolant flow rate. If the flow rate is turbulent [9] and the conduction term is neglected, the coefficient is approximated by

$$h = kw^{0.8}, \tag{4.32}$$

where k is a geometrically dependent constant.† For a nuclear rocket test reactor it was found that $k \approx c_n/(n^{0.8}d^{1.8})$, where n is the number of coolant passages (cylindrical) and d is the average passage diameter; c_n was approximately $6.2(10)^{-3}$ Btu/(in. sec)$^{0.2}$ °R lb$^{0.8}$ in this case.

The remaining coefficients (c_{m}, c_{g}, c_{mg}) usually have a weak temperature dependence over a broad range of high temperatures. Therefore, if (4.32) is approximated by a linear function, $h = kw$, the process is bilinear with w and q as input variables.

† More accurate descriptions of heat transfer coefficients are presented by Humble *et al.* [10].

The model frequently may be further simplified by neglecting coolant mass heat capacity and coolant temperature at the core inlet. Then Eqs. (4.29) and (4.30) may be approximated by

$$dT_u/dt = q/c_{\mathrm{m}} - awT_u, \tag{4.33}$$

where a is a heat transfer constant.

Suppose it is assumed for the present that $q(t)$ and $w(t)$ are at least piecewise continuous controls with upper and lower physical constraints so that

$$0 \leq q(t) \leq q_b$$

and

$$w_a \leq w \leq w_b.$$

If it is desired to raise the average core temperature in minimum time or with minimum coolant, it is obvious from the maximum principle or physical reasoning that

$$q(t) = q_b$$

and that

$$w(t) = w_a \qquad \text{on} \quad (t_0, t_{\mathrm{f}}).$$

At the endpoints $T_u(t_0) = T_{u0}$ and $T_u(t_{\mathrm{f}}) = T_{u\mathrm{f}}$ with $dT_u/dt = 0$. Therefore,

$$q(t_0) = ac_{\mathrm{m}}T_{u0}w(t_0) \qquad \text{and} \qquad q(t_{\mathrm{f}}) = ac_{\mathrm{m}}T_{u\mathrm{f}}w(t_{\mathrm{f}}).$$

4.3 REACTOR CONTROL

The neutronic and heat exchange state equations (4.6), (4.7), and (4.33) are coupled together in a physical reactor by coefficients of reactivity. The most common reactivity coefficient couples core temperature T_u and neutron level n or power q by means of a temperature reactivity† which is commonly approximated by

$$\delta k_t = \alpha_t T_u, \tag{4.34}$$

where α_t usually may be assumed constant within the accuracy of which it may be estimated. The temperature coefficient is usually negative as a consequence of thermal expansion which reduces core density, and

† Ash [11, pp. 79–86] gives a detailed analysis of temperature reactivity.

thereby neutron moderation and reactivity. Also, core temperature affects neutron capture cross sections, fission cross sections, absorption cross sections, and so-called Doppler broadening associated with the resonance region of the neutron energy spectrum [12, p. 180]. For certain fast reactors, the Doppler coefficient is positive and so large that the overall temperature coefficient can be positive. Fortunately, this is not usually the case. For KIWI A prime, one of the first nuclear rocket test reactors, $\alpha_t \approx -1.52(10)^{-5}/°\text{R}$.

A negative coefficient of reactivity almost always has a stabilizing effect on the reactor dynamics. Even for a linear heat exchange model (h = constant), however, the multiplicative term in the neutron kinetics, (4.4) and (4.5), causes the reactor to be nonlinear and indeed nonbilinear.

In addition to temperature–reactivity feedback, some reactors use a coolant which is also a neutron moderator. For example, liquid hydrogen propellant generally used with nuclear rockets is a very good moderator. Even as a dense gas at the reactor core entrance, it has a strong moderating effect on intermediate energy (thermal) neutrons and on fast neutrons. Such coolant reactivity is approximated by

$$\delta k_p = k_p \rho_g V, \tag{4.35}$$

where k_p is a positive constant, V is the coolant volume in the core, and ρ_g is the coolant density. While this reactivity feedback is positive, it is apparent that again the temperature effect is negative according to the gas equation,

$$\rho_g = \frac{p}{zRT_g}, \tag{4.36}$$

where p is the average coolant pressure in the core, R is the universal gas constant, and z is the compressibility factor (usually of value 1). The coolant density reactivity also couples the neutron kinetics to the coolant feed system.

The total reactivity δk, which is the input to the neutron kinetics, (4.4) and (4.5), is a summation of reactivity, described by

$$\delta k = \delta k_t + \delta k_p + \delta k_c, \tag{4.37}$$

where δk_c is the reactivity generated usually by neutron absorbers or reflectors in the form of control rods or control vanes. Since meaningful safety constraints are put on δk rather than δk_c, it is assumed here that the admissible control must satisfy $|\delta k| \leq \gamma$. (Symmetrical constraints are used for convenience.) This has the effect of decoupling the control loops, as is analyzed next.

Suppose the reactor dynamics are described by the nuclear fission process, (4.4) and (4.5), with $q(t)$ proportional to $n(t)$, and by the simple heat exchange process (4.33).† Let the control variables be $u_1 = \delta k(t)$ and $u_2 = w(t)$, and combine constants in (4.35) and (4.36) so that

$$k_p = \frac{c_p u_2}{(T_g)^{1/2}} \tag{4.38}$$

where c_p is a positive constant.

4.3.1 Maximum Principle Solution

Now the problem is to start the system on minimal coolant ($f_0 = u_2$) from some initial equilibrium state with $q(t_0) = q_0$, $T_u(t_0) = T_{u0}$, and $u_2(t_0) = u_a$ to some terminal condition so that $q(t_f) = q_f$, $T_u(t_f) = T_{uf}$. Also, $\dot{q}(t) = 0$ and $\dot{T}_u(t) = 0$ for $t > t_f$. Control and state constraints include

$$|u_1(t)| \leq \gamma, \tag{4.39}$$

$$w_a \leq u_2(t) \leq w_b, \tag{4.40}$$

and

$$q(t) \leq q_f. \tag{4.41}$$

Reasons for the control constraints have been discussed, and the state constraint on power may be a maximum attainable power for the critical assembly or a safety constraint on power overshoot.

It is apparent from the maximum principle that the optimal control while $\mathbf{x}(t)$ is not on a state constraint boundary is described again by a bang–bang process with

$$u_1 = \operatorname{sgn} p_1(t) \tag{4.42}$$

† Due to the homogeneity of the fission process, $q(t)$ may be written in place of $n(t)$ in (4.4) with $c_i(t)$ $(i = 1, \ldots, 6)$ assumed to have equivalent units of power in (4.4) and (4.5).

$$u_2 = \begin{cases} w_a & \text{for} \quad S(\mathbf{x}, \tilde{\mathbf{p}}) < 0 \\ w_b & \text{for} \quad S(\mathbf{x}, \tilde{\mathbf{p}}) > 0, \end{cases} \tag{4.43}$$

where

$$S(\mathbf{x}, \tilde{\mathbf{p}}) = p_0 - ap_8T_u, \tag{4.44}$$

and $\tilde{\mathbf{p}}$ is described again by

$$d\tilde{\mathbf{p}}/dt = -\partial H/\partial \tilde{\mathbf{x}}, \tag{4.45}$$

TABLE 4.2

Design Conditions for Hypothetical Rocket Engine[a]

Thrust	100,000 lb
Specific impulse (including losses)	760 sec
Reactor power	2260 MW
Propellant weight flow rate	130 lb/sec
Nozzle entrance stagnation pressure	1100 psia
Reactor exit temperature	4500°R
Reflector entrance pressure	1220 psia
Nozzle exit pressure	4.0 psia
Nozzle throat area	61 in.^{2}
Nozzle expansion ratio	20
Gas temperature at reflector entrance	120°R
Gas temperature at core entrance	220°R
Neutron mean effective generation time	1.4×10^{-5} sec
Propellant reactivity	0.0280
Temperature reactivity	−0.0073
Heat exchange thermal time constant	1.5 sec
Control actuator bandwidth and damping	50 sec^{-1}, 0.7
Turbopump speed, closed-loop bandwidth	40 sec^{-1}

Controller design (proportional plus integral, $k_p + k_1/s$)

Thrust control

$k_{pF} = 0.001$ sec^{-1} lb^{-1}, $k_{1F} = 0.010$ sec^{-2} lb^{-1}

Temperature control

$k_{pT} = 0.009$ ($\log_e$ Btu/sec)/°R, $k_{1T} = 0.040$ ($\log_e$ Btu/sec)/°R-sec

Power control

$k_{pp} = 0.640$ 1/$\log_e$ Btu/sec, $k_{1p} = 2.00$ 1/sec-$\log_e$ Btu/sec

Closed loop bandwidths

Thrust: 25 sec^{-1}; temperature: 20 sec^{-1}; power: 50 sec^{-1}

[a] See Mohler and Shen [6].

TABLE 4.3

OPTIMAL REACTOR STARTUP TRAJECTORY WITH $q(t) \leq q_f$ [6][a]

Time, t (sec)	Power, q (10^4 Btu/sec)	Temperature, T_e (10^2 °R)	$p_1(t)$ (10^{-6} sec²/Btu)	$p_8(t)$ (10^{-4} sec/°R)
0	1.3144	5.0000		
$t_0 = 0.1$	1.3144	5.0000	11.946	7.2139
0.2	13.253	5.0651	9.2838	7.2306
0.4	24.889	5.33720	5.6093	7.2640
0.6	42.642	5.9257	3.3932	7.2976
0.8	71.793	6.8790	2.0589	7.3313
1.0	119.91	8.4903	1.2663	7.3652
$t_a = 1.24$	214.20	11.911	5.0718	7.4101
1.4	214.20	14.868	4.9676	7.4375
1.6	214.20	18.549	4.8368	7.4718
1.8	214.20	22.213	4.7054	7.5064
2.0	214.20	25.860	4.5734	7.5411
2.2	214.20	29.490	4.4408	7.5759
2.4	214.20	33.104	4.3076	7.6110
2.6	214.20	36.701	4.1737	7.6461
2.8	214.20	40.281	4.0393	7.6815
$t_1 = 3.08$	214.20	45.011	3.8500	7.7751

[a] Input data: $u_1 = 0$, $t < 0.1$ sec $= t_0$; $u_1 = 0.9\beta$, $t \geq 0.1$, $p_{10} > 0$, and $q < q_1 = 2.142(10)^6$ Btu/sec, for $t < t_a$; $u_1 = l\sum_{i=1}^{6} \dot{c}_i/q_1$, $q = q_1$, for $t > t_a$; $u_2 = u_a = 2$ lb/sec, $T_u < T_{uf}$, and $S < 0$, for $t > t_1$; $a = 1.153 \times 10^{-2}$, $T_u(0) = 500$°R; $dT_u(t_1)/dt = dq(t_1)/dt = 0$. Other data are given in Tables 4.1 and 4.2.

with $H(\mathbf{x}, \tilde{\mathbf{p}}, \mathbf{u}) = \langle \tilde{\mathbf{p}}, \mathbf{f}(\mathbf{x}, \mathbf{u}) \rangle$ [7]. Since there is a constraint on total reactivity, $u_1(t)$, no increase in coolant flow rate, $u_2(t)$, would permit a faster change of state nor a consumption of less coolant than does $u_1(t) = \gamma$ and $u_2(t) = w_a$. At the power constraint boundary, however, it is again necessary that $u_1 = l\sum_{i=1}^{6} \dot{c}_i/q_f$. Again, this control is unstable in practice, but may be approximated by a simple closed-loop control or a dither control. At the desired terminal temperature, flow rate is increased so that $u_2(t_f) = q_f/c_m a T_{uf}$. For the nuclear rocket, this corresponds to the designed thrust and temperature condition. Figure 4.8 shows the minimal propellant control of a rocket reactor whose design parameters are given in Tables 4.1 and 4.2. It may readily be seen that these trajectories satisfy the maximum principle and further are indeed optimal trajectories [6].

Table 4.3 quantifies pertinent state and costate values during the coolant-optimal reactor start shown in Fig. 4.8. The costate equations are defined by (4.43) while $q(t) < q_f$ and $t_0 \leq t < t_a$. On the interval (t_a, t_f), $q(t) = q_f$ and the costate is defined by

$$d\tilde{\mathbf{p}}/dt = -\partial H/\partial\tilde{\mathbf{x}} + \rho\, \partial z/\partial\tilde{\mathbf{x}}, \tag{4.46}$$

where $z(\tilde{\mathbf{x}}, \mathbf{u}) = \dot{q}(\tilde{\mathbf{x}}, \mathbf{u})$. Also, with $u_2(t) = w_a$ on (t_a, t_f), it is necessary from the addended maximum principle [7] that

$$\partial H/\partial\mathbf{u} = \rho\, \partial z/\partial\mathbf{u} + \gamma\, \partial r/\partial\mathbf{u}, \tag{4.47}$$

where

$$r = u_a - u_2. \tag{4.48}$$

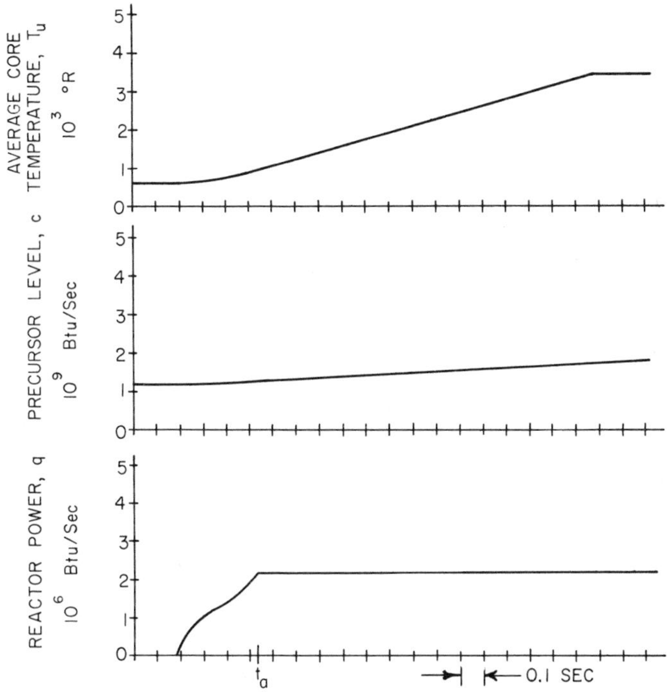

Fig. 4.8. Minimal propellant control of a nuclear rocket [6].

Further, it is seen that

$$\gamma = S(t) = (p_0 - ap_8T_u),$$

and

$$\rho = p_1(t).$$

Since $T_u(t_f) = T_{uf}$ and $q(t_f) = q_f$, the transversability condition requires that $p_j(t_f) = 0, j = 2, \ldots, 7$.

At the constraint entrance corner, that is, when $t = t_a$, $\mathbf{p}(t)$ may be assumed continuous since there is no exit corner [7]. Again, it is readily seen that the problem cannot have a nondegenerate singular solution.

4.3.2 Quasi-linear Programming Solution

Consider the previous problem with the fission process approximated by one average precursor group, that is, the state $x_1 = n(t)$ or $q(t)$, $x_2 = c(t)$, and $x_3 = T_u(t)$ is defined by (4.11) and (4.33) with control magnitude constraints, (4.39) and (4.40), along with control rate constraint

$$|du_1/dt| \leq \eta. \tag{4.49}$$

Further, it is assumed that $q_l \leq q(t) \leq q_m$, $T_l \leq T_u \leq T_m$, and $T_l' \leq dT_u/dt \leq T_m'$. Of course, all the state variables must be positive values of power, equivalent precursor power, and absolute temperature. The inequalities may be put into the form of equalities for solution by the simplex algorithm by the addition of so-called slack variables [6]. To solve the optimal control problem by linear programming as discussed in Chapter III, the nonlinear state equations may be approximated about some "optimal" trajectory (designated by an asterisk) by linear algebraic difference equations. Then the problem is solved sequentially. For the ith iteration, the difference equations for the single-precursor reactor dynamics may have the form of

$$\left[1 - \frac{(u_{1i}^* - \beta)\Delta t^*}{2l}\right] q_{i+1} - \left[1 + \frac{(u_{1i}^* - \beta)\Delta t^*}{2l}\right] q_i - \frac{\lambda \Delta t^*}{2}(c_{i+1} + c_i)$$

$$- \frac{u_{1i}\Delta t^*}{2}(q_{i+1}^* + q_i^*) + \left[\left(\frac{u_{1i}^* - \beta}{l}\right)(q_{i+1}^* + q_i^*) + \lambda(c_{i+1}^* + c_i^*)\right]\frac{\Delta t}{2}$$

$$= \left[\frac{\beta}{l}(q_{i+1}^* + q_i^*) - \lambda(c_{i+1}^* + c_i^*)\right]\frac{\Delta t^*}{2},$$

$$\left(1+\frac{\lambda\Delta t^*}{2}\right)c_{i+1}-\left(1-\frac{\lambda\Delta t^*}{2}\right)c_i-\frac{\beta\Delta t^*}{2}(q_{i+1}+q_i)$$

$$-\left[\frac{\beta(q_{i+1}^*+q_i^*)}{l}-\lambda(c_{i+1}^*+c_i^*)\right]\frac{(\Delta t-\Delta t^*)}{2}=0,$$

and

$$\left(1+\frac{au_{2i}^*\Delta t^*}{2}\right)T_{i+1}-\left(1-\frac{au_{2i}^*\Delta t^*}{2}\right)T_i-\frac{\Delta t^*}{2c_{\mathrm{m}}}(q_{i+1}+q_i)$$

$$+\frac{a\Delta t^*(T_{i+1}^*+T_i^*)}{2}u_{2i}$$

$$-\left[\frac{q_{i+1}^*+q_i^*}{c_{\mathrm{m}}}-au_{2i}^*(T_{i+1}^*+T_i^*)\right]\frac{\Delta t+\Delta t^*}{2}=0,$$

where $\sum_{i=1}^{N}\Delta t_i = t_{\mathrm{f}}$, and N is the number of increments. Then the performance index may be approximated by

$$J=\sum_{i=1}^{N}(u_{2i}^*\Delta t+u_{2i}\Delta t^*-u_{2i}^*\Delta t^*),$$

which is the negative of the classic objective function used in linear programming.

For the following computations, 50 time increments ($N = 50$) were used. Here, the necessary work matrix was generated by a FORTRAN program so that the data were put in a form compatible with MPS/360. Then the output (linear programming solution) from MPS/360 is used as input to the same FORTRAN program to a new work matrix. For each new linear programming sequence, the previous "optimal" solution is used to start the iterative solution. Convergence is assumed when any linear programming solution is within the standard tolerance of MPS/360.

The minimal propellant startups shown in Figs. 4.9 to 4.11 again utilize the nuclear rocket parameters given by Table 4.2 unless otherwise noted, and the inequality constraints are given by $\gamma = 0.9\beta$, $w_a = 2$ lb/sec, $w_b = 400$ lb/sec, $q_1 = 0.01$ MW, $T_l = 0$°R, and $T_m =$ 4500°R, and η, q_m, and T_m' take on various values as indicated on the figures. The T_l' does not enter into the computations.

The first computation, presented in Fig. 4.9, was made in the absence

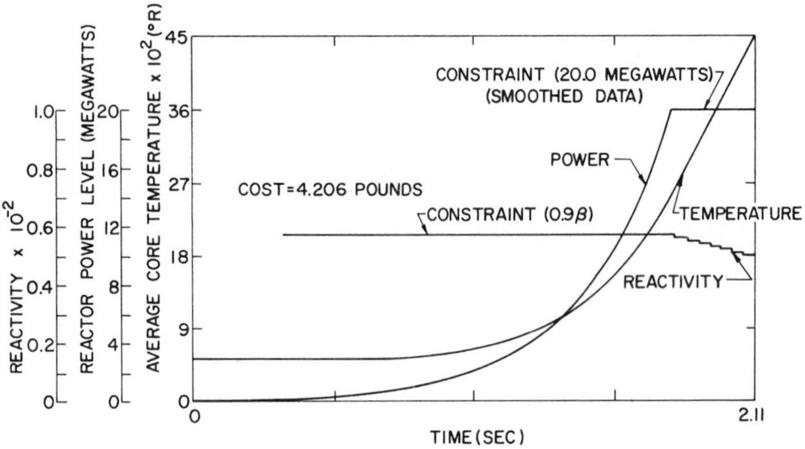

Fig. 4.9. Minimal propellant rocket start with power constraint [$q(t) \leq$ 20 MW] and propellant flow at low constraint [6].

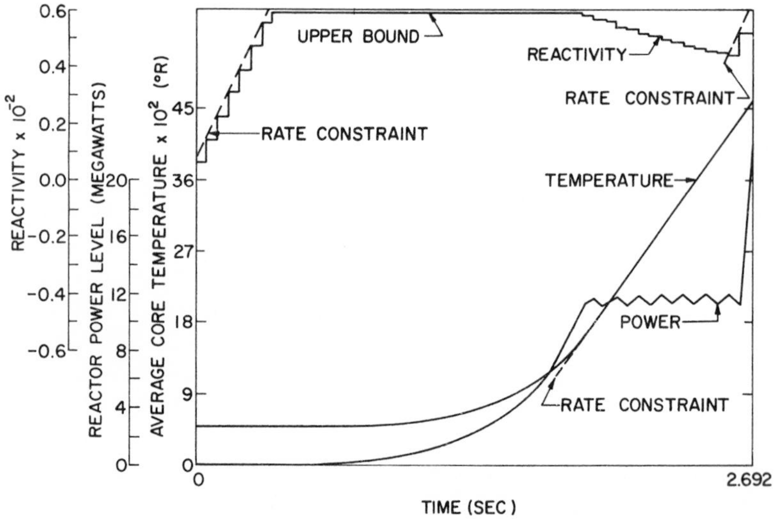

Fig. 4.10. Minimal propellant rocket start with temperature rate constraint ($dT_u/dt \leq 3600$°F/sec), reactivity rate constraint ($|d\delta k/dt| \leq 0.015$ sec^{-1}), and propellant flow at low constraint except on last interval where $u_2 = 19.42$ lb [6].

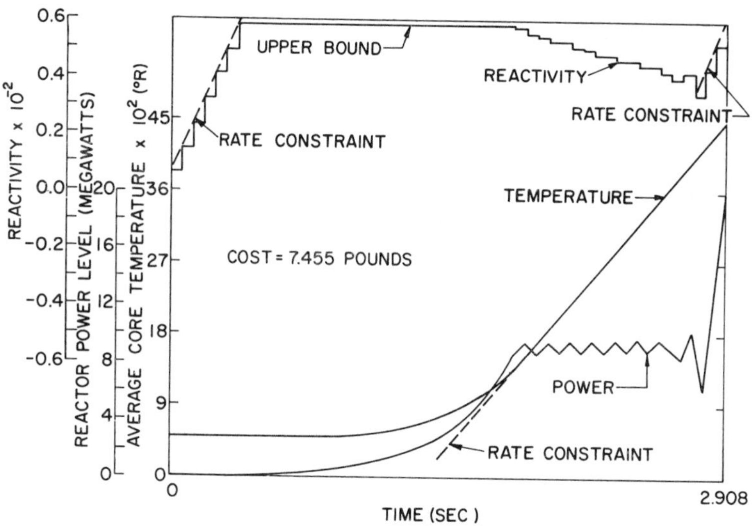

Fig. 4.11. Minimal propellant rocket start with temperature rate constraint [$T_u(t) \le 2880°\text{R/sec}$], reactivity rate constraint ($|d\delta k/dt| \le 0.015\ \text{sec}^{-1}$), and propellant flow at low constraint except on last interval where $u_2 = 30.17$ lb [6].

of reactivity rate constraints and temperature rate constraints. Here the starting point for the first iteration was obtained by integrating the system equations for 1.5 sec with reactivity on its upper bound and propellant flow rate on its lower bound, as suggested by the analysis made in the previous section. Obviously, the process is again one of minimal time as well as minimal fuel.

The optimal policy is as analyzed in the previous section. That is, maximum reactivity is added to the reactor until power reaches its constraint, at which time reactivity must again maintain $dq(t)/dt = 0$. During the entire startup, propellant flow rate is at its minimum allowable value. Then at the terminal time, flow rate must be increased to maintain the desired terminal temperature according to $u_2 = q_f/c_m a T_{uf}$. For those problems, it is assumed that the terminal temperature $T_{uf} = T_m = 4500°\text{R}$.

Optimal startups of the nuclear rocket reactor with rate constraints imposed on reactivity and temperature are shown in Figs. 4.10 and 4.11. The starting iterations for Fig. 4.10 used a strategy that is similar to that for Fig. 4.9. Then for the subsequent figures the optimal control from the previous problem is used to start the computation.

The computation times (for an IBM 360/40) and number of linear programming iterations are listed in Table 4.4 for the startups graphed in Figs. 4.9 to 4.11. The two computation times shown in the table correspond to: (1) the time to arrive at an "optimal" solution to each separate linear programming problem after the basic feasible solution

TABLE 4.4

Nuclear Rocket Engine MPS/360 Computation Times[a]

Figure	Number of iterations to feasible	Number of iterations to optimal[b]	Major iteration	Computing time (min)[c]
4.9	187	—	1	3.13
	—	70	1	8.16
	Feasible at start	—	2	2.62
	—	2	2	2.79
	Feasible and "optimal" at start			2.60
			Total	19.50
4.10	21	—	1	5.63
	—	40	1	3.03
	34	—	2	10.42
	—	29	2	7.25
	16	—	3	4.25
	—	14	3	2.92
	19	—	4	5.05
	—	17	4	4.38
	Feasible and "optimal" at start			2.53
			Total	45.46
4.11	9	—	1	3.94
	—	8	1	2.81
	2	—	2	2.42
	—	4	2	2.09
	Fesible and "optimal" at start			2.62
			Total	13.88

[a] See Mohler and Shen [6].

[b] "Optimal" here refers to optimal for the linear program.

[c] These computation times do not include running of the FORTRAN program between each major iteration. One run of this program requires approximately 1.50 min. Time refers to IBM 360/40.

has been obtained, and (2) the time to arrive at a basic feasible solution to the constraint equations.

The factors that affect computation time most include the number of constraints, the density of the work matrix, the number of variables, and the number of time increments.

From Figs. 4.10 and 4.11, instability of the constraint control, $u_1 = lc/q_m$, is apparent due to the error associated with the difference equation approximation.

Optimality of these trajectories is at least plausible from physical arguments, for the desired terminal state is reached in the minimum time allowed by the system constraints. Again, reactivity is inserted at its maximum rate until its upper magnitude constraint is reached. Meanwhile, reactor power increases in minimal time until the average core temperature T_u reaches its rate constraint. Then reactivity varies so that the necessary power change maintains $dT_u/dt = 2880°\text{R/sec}$ in Eq. (4.33). Notice from the figure that the required power is nearly constant. However, such would not be the case at larger flow rates. Just prior to the terminal time, with $T_u(t_f) = 4500°\text{F}$, reactivity must increase at its maximum rate, so that reactor power increases in minimal time to its terminal state [here, $q(t_f) = 20$ MW]. Also, coolant flow rate u_2 must increase when reactivity increases to maintain the dT_u/dt constraint of 2880°R/sec. Though not shown on the graphs, u_2 must increase to $q(t_f)/ac_m$ at the terminal time to maintain constant temperature according to Eq. (4.33), and reactivity u_1 must hold power constant while maintaining continuity at the terminal time.

While it might appear that the problem has a singular solution, it is shown by Mohler and Shen [6] that no normal singular solution exists.

4.3.3 *Suboptimal Closed-Loop Control*

Simple closed-loop control with suboptimal performance is desirable for most applications. Such control has been studied for the nuclear fission process in Section 4.1.4. A classical nuclear engine control system such as that used for static tests includes feedback of temperature, power, and thrust or pressure. A simpler, but poorer performing, system controls on only one output such as thrust, but depends on coolant reactivity calibration. Both are discussed in detail by Mohler and Perry [13]. The latter control utilizes the propellant density reactivity to adjust power. High-performance controller gains for the former system are given in Table 4.2. Here, proportional plus integral controls are used with gains of k_p and k_i, respectively.

It should be realized that only simplified models of the actual reactor systems are analyzed in this chapter. For example, several core heat exchange lumps with appropriate reactivity feedbacks are usually utilized in practical controller design analysis. The nuclear rocket engine includes nozzle and reflector heat exchangers, a complicated propellant feed system, and as for any such system, transducers and control actuators. The feed system may involve a complicated turbine-driven pump, which usually is powered by hot bleed gas obtained from the reactor, fluid compressibility, and transport delay times [6]. Turbopump constraints due to stall and cavitation, which are defined on performance maps by lines of constant specific speed, have been neglected in this analysis [14]. For example, it may be necessary to increase propellant (coolant) flow rate more than the previously considered constraints required so that the pump does not stall [6].

As an example of practical system complication, it might be noted that one dynamical model utilized to analyze the control of the nuclear rocket engine included about 40 state equations and many more algebraic equations. Typically, the constraints and equations are nonlinear and nonconvex. Bilinear models, however, arise in a natural manner in the derivation of many of the state equations. Besides the heat exchange and nuclear fission processes, the turbopump has a bilinear control term arising from a multiplication of speed control valve position and turbine inlet pressure [13].

4.4 REACTOR SHUTDOWN

Time-optimal, coolant-optimal, and suboptimal changes in reactor state have been analyzed in Section 4.3. As mentioned in Section 4.1, however, for large sudden decreases in flux there is a slowly decaying fission heat that is not accounted for by the previous model. Fortunately, this afterheat can be added to the model with the power attenuation determined by the decay of the radioactive fission products [15].

For many power reactors there is an even further complication due to the accumulation of poison buildups which may offer severe constraints to reactor restarts after shutdown. Two alternatives are to let the poison concentration decay sufficiently or to provide sufficient reactivity to override the poison. To minimize the losses due to excessive shutdown periods, optimal control can play a significant role.

4.4.1 Bilinear Poison Dynamics

Similar to other regenerative processes, poison concentration may be described by a bilinear model. In this case

$$dx_1/dt = \gamma_1 u - \lambda_1 x_1 \tag{4.50}$$

and

$$dx_2/dt = \gamma_2 u + \lambda_1 x_1 - \lambda_2 x_2 - \sigma_2 u x_2, \tag{4.51}$$

where $0 \le u(t) \le u_b$, $x_1(t) = [^{135}\mathrm{I}]/\Sigma_\mathrm{f}$, $x_2(t) = [^{135}\mathrm{Xe}]/\Sigma_\mathrm{f}$; u is the thermal neutron flux; [I] is the iodine concentration; [Xe] is the xenon concentration; Σ_f is the macroscopic fission cross section; γ_1 is the fractional yield of ^{135}I equal to 0.061; γ_2 is the fractional yield of ^{135}Xe equal to 0.002; λ_1 is the ^{135}I decay constant equal to $2.9(10)^{-5}$ sec^{-1}; λ_2 is the decay constant of ^{135}Xe equal to $2.1(10)^{-5}$ sec^{-1}; and σ_2 is the microscopic capture cross section of ^{135}Xe for thermal neutrons equal to $2.7(10)^{-18}$ cm^2. The dynamics of this process are so slow that neutron flux can be used as a control variable rather than reactivity.

4.4.2 Time-Optimal Shutdown

A sensible optimal control problem here is to change states in minimum time from $\mathbf{x}(t_0) = \mathbf{x}_0$ to $\mathbf{x}(t_\mathrm{f})$ that belongs to some terminal set Γ_f so that xenon concentration is constrained by $x_2 \le x_{2m}$. This analysis follows that of Roberts and Smith [16]. The equilibrium set for equilibrium flux values on the xenon–iodine state plane are shown in Fig. 4.12. Here, $x_{2m} = 5(10)^{16}$ cm^{-2}, and the terminal set Γ_f is the zero-flux shutdown line, which does not violate this constraint, but which just reaches it.

From the maximum principle, it is obvious that, while the state is not on the constraint boundary, the extremal control is bang–bang with

$$u = \begin{cases} 0 & \text{for } \; S < 0 \\ u_b & \text{for } \; S > 0, \end{cases} \tag{4.52}$$

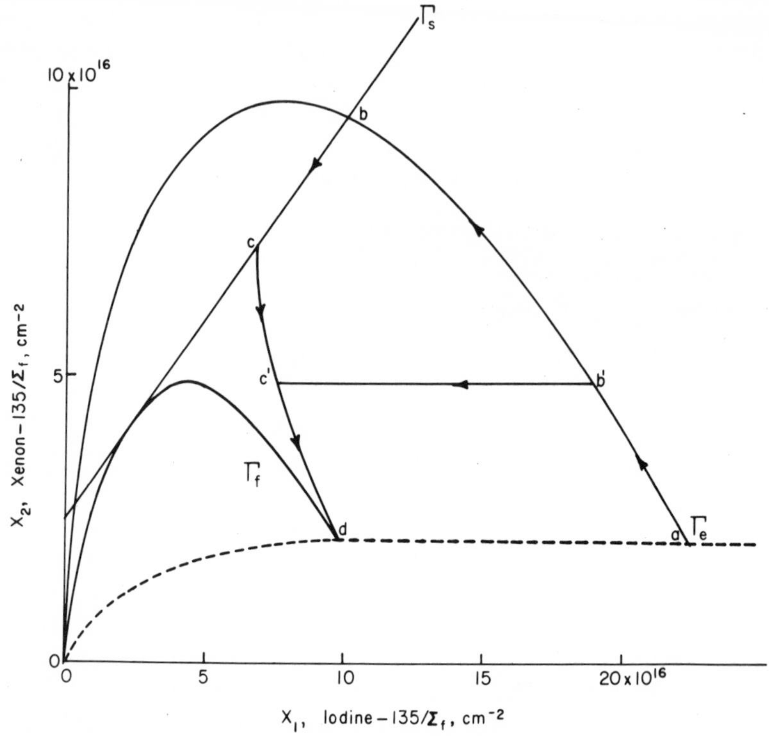

Fig. 4.12. Reactor shutdown trajectories in iodine–xenon state plane [16].

where the switching function is

$$S = -\gamma_1 p_1 + p_2(\sigma_2 x_2 - \gamma_2) \tag{4.53}$$

and $\mathbf{p}(t)$ is described by the usual adjoint equations of (4.50) and (4.51).

From the state equation, the target set Γ_t can be described by

$$x_2 = g(x_1) = \left[\frac{x_{2m}}{(\alpha x_{2m})^\alpha} + \left(\frac{\alpha x_{2m}}{1-\alpha}\right)^{1-\alpha}\right] x_1{}^\alpha - \frac{x_1}{1-\alpha}, \tag{4.54}$$

where $\alpha = \lambda_2/\lambda_1$. Then the orthogonal transversality condition requires that $p_1 + p_2\, dg/dx_1 = 0$, where

$$\frac{dg}{dx_1} = \alpha\left[\frac{x_{2m}}{(\alpha x_{2m})^\alpha} + \left(\frac{\alpha x_{2m}}{1-\alpha}\right)^{1-\alpha}\right] x_1^{\alpha-1} - \frac{1}{1-\alpha}.$$

For this problem, it is possible for the switching function $S(t)$ to vanish on some finite time interval. In such a case, the control is singular, and must only force the state to satisfy $S(t) = 0$ and $dS/dt = 0$. From these conditions, a singular set Γ_s may be computed so that whenever trajectories intersect such a set, the singular solution must be considered. Hence, Γ_s is defined by

$$x_2 = x_1 + \left(\frac{\gamma_1 + \gamma_2}{\sigma_2} - \frac{\gamma_2\lambda_2}{\sigma_2\lambda_1}\right) = x_1 + 2.28(10)^{16}\ \mathrm{cm}^{-2}. \quad (4.55)$$

This singular set is graphed in Fig. 4.12. It is shown that if the [Xe] constraint were relaxed, the optimal trajectory *a–b–c–d* involves a singular control on the *b–c* arc, where point *a* corresponds to equilibrium with $u = 10^{14}$ neutrons/(cm^2 sec).

The optimal trajectory is computed from the equations above and involves $u = 0$ on *a–b*, $u = u_b = 10^{14}$ neutrons/(cm^2 sec) on *c–d*, and, for the singular portion,

$$u = \frac{x_1(2\lambda_1 - \lambda_2) - \lambda_2 h}{\sigma_2 x_1 + 2\gamma_1 - (\lambda_2\gamma_2/\gamma_1)}$$

on *b–c*, where from Eq. (4.55), $h = 2.28(10)^6\ \mathrm{cm}^{-2}$. [It can be shown readily that $u(t)$ on *b–c* is less than u_b.]

With the state constraint on x_2, however, the flux to maintain $x_2(t) = x_{2m}$ is computed from Eq. (4.50). Thus,

$$u = \frac{\gamma_2 x_{2m} - \lambda_1 x_1}{\gamma_2 - \sigma_2 x_{2m}}, \quad (4.56)$$

which also is less than u_b.

If the constraint $x_2 \le x_{2m} = 5(10)^{16}$ is imposed, however, the costate must satisfy the addended adjoint equation (4.46) on the boundary. At the constraint entrance corner, the costate may have a jump discontinuity with

$$\mathbf{p}(t_a^-) = \mathbf{p}(t_a^+) + \mu \left.\frac{\partial g}{\partial \mathbf{x}}\right|_{t_a},$$

where $g(\mathbf{x}) = x_2 - x_{2m} = 0$. Again, continuity may be assumed at the exit corner.

From these necessary conditions, it is not too difficult to show that singular control is not needed with such a low x_{2m}, and the trajectory

a–b'–c'–d shown in Fig. 4.12 is optimal with the following policy: $u = 0$ on a–b', $u = u_m$ on c'–d; and $u(t)$ computed by (4.56) on b'–c'. At point d, u should return to zero to maintain the desired terminal set. The time consumed in traversing the trajectory a–b'–c'–d is 865 min [16]. With such a slow response, the neutron response time for reactivity changes is certainly negligible.

4.4.3 Suboptimal Shutdown for Minimax Xenon

In some cases, it might be more desirable to minimize the peak xenon buildup rather than accept some maximum x_{2m} and to minimize time as done in the last section. This is the formulation used by Rosztoczy and Weaver [17] and Ash [18], where the problem is solved by the maximum principle and dynamic programming, respectively. It should be cautioned, however, that in both of these papers, it is assumed that singular solutions cannot affect the optimal shutdown. As was shown in the previous case with the xenon constraint relaxed, singular subarcs can form an important part of the optimal shutdown.

Following the analysis of Rosztoczy and Weaver [17], the peak xenon concentration reaches after shutdown is related to iodine and xenon concentrations at the time of shutdown t_f with $u = 0$, and $dx_2/dt = 0$ in Eq. (4.51). The time of peak xenon concentration occurs at

$$t - t_f = \frac{1}{\lambda_1 - \lambda_2} \log_e \frac{\lambda_1\lambda_2(x_{1f} + x_{2f}) - \lambda_2{}^2 x_{2f}}{\lambda_1{}^2 x_{1f}}.$$

Then, from the solutions to Eqs. (4.50) and (4.51), the maximum xenon concentration is related to $\mathbf{x}(t_f) = \mathbf{x}_f$ by

$$x_{2m} - (x_{11} + a_1 x_{21})(a_2 + a_1 a_2\, x_{21}/x_{11})^{a_3},$$

where $a_1 = 1 - (\lambda_2/\lambda_1)$, $a_2 = \lambda_2/\lambda_1$, and $a_3 = 2/(\lambda_1 - \lambda_2)$.

Now the problem can be formulated in terms of a terminal index of performance

$$J(x_1) = x_{2m} = (x_{11} + a_1 x_{21})(a_2 + a_1 a_2\, x_{21}/x_{11})^{a_3}.$$

As before, the problem is to find the admissible control $0 \leq u(t) \leq u_b$ which minimizes $x_0(t_f) = J$. This terminal index may be expressed in terms of an integral index

$$J = \int_{t_0}^{t_f} f_0(\mathbf{x}, \mathbf{u})\, dt,$$

where

$$f_0 = \langle \partial J(\mathbf{x})/\partial \mathbf{x}, d\mathbf{x}/dt \rangle.$$

Again, $H = \langle \tilde{\mathbf{p}}, \tilde{\mathbf{f}}(\mathbf{x}, \mathbf{u}) \rangle$ is linear in control, and it is maximized by a bang–bang control if the singular solution with a zero switching function is neglected.

By trial and error, the switching times are computed by Rosztoczy and Weaver [17] for assumed numbers of switchings. For example, the suboptimal two-pulse policy results in a decrease of maximum xenon by about 16 percent from that of the one-pulse policy. Also, the two-pulse policy shows that the maximum xenon buildup can occur for $t < t_f$, and for some applications, the problem formulation should be altered. Since the control is suboptimal in nature anyway, the control might be handled as one of state constraint $x_2(t) \leq x_{2m}$ on $[t_0, t_f]$. (In practice, the state constraint control and singular control can sometimes be approximated by a multiple on–off control.)

4.5 CONCLUSIONS

Mathematical models of a bilinear form are derived in a natural manner to approximate the description of nuclear fission and heat transfer. Again, the failure of linear models is quite evident. Secondary effects give rise to coefficients that are functions of state, but the most significant nonlinearity results from the multiplication of state and control in the dynamical process. For nuclear fission, bilinear control arises as a consequence of neutron level or power (a state variable) being multiplied by reactivity or neutron multiplication (a control variable). A multiplication of coolant flow rate (a control variable) and temperature (a state variable) is produced in heat transfer between a solid wall, such as a reactor core, and moving coolant fluid. Even the generation of poison products in nuclear reactors may be described by a bilinear model with thermal neutron flux (the control) multiplying xenon concentration.

Again, as for the optimal bilinear control processes with the linear-in-control performance indices discussed in Chapter III, bang–bang control is frequently the best possible process in the absence of state and control rate constraints. The latter are shown to play an important role for some reactor control problems, and in this regard the quasilinear programming algorithm, which has been introduced in Chapter III, is shown to be effective. Also, it is indicated that suboptimal closed-loop controllers may be designed for reactor control.

Exercises

4.1 In some cases, it is assumed that all neutrons are generated promptly. With this assumption, calculate the time-optimal reactivity and neutron level to control the reactor fission process with $l = 3(10)^{-4}$ sec from $n(0) = n_0$ to $n(t_f) = 100n_0$ with

(a) $|u| \leq \gamma$
(b) $|u| \leq \gamma$ and $|\dot{u}| \leq \eta$

where γ and η are positive constants.

4.2 Repeat Exercise 4.1 for a reactor fission process with all precursors lost to the system.

4.3 Methods have been suggested to control a reactor fission process by means of a manipulated neutron source. Compute the time-optimal source control and neutron level to control the single precursor process with $l = 10^{-5}$ sec and reactivity $u = 0.5\beta$ from an equilibrium state with $n(0) = n_0$ to $n(t_f) = 1000n_0$ so that source $|s| \leq s_a$, a positive constant.

4.4 For the single-precursor kinetics, what is the required terminal control to maintain $\dot{n}(t) = 0$ with $\dot{c}(t) \neq 0$ for $t > t_f$? Is this control admissible?

4.5 Show that the terminal reactivity control described by (4.20) is admissible.

4.6 (a) How many Volterra kernels are required to describe the single-precursor nuclear fission process (4.6) and (4.7)?
(b) Compute the precise form of the required kernels.
(c) Provide the precise form of Volterra series for a specified $u(t)$.

4.7 Discuss the role of singular solutions in optimal reactor startup and shutdown.

4.8 Analyze controllability of reactor shutdown and restart with respect to xenon and iodine poisons.

4.9 Does the neutron process (4.2) and (4.3) with source control satisfy the sufficiency condition for complete controllability given in Chapter III? Discuss your answer and the controllability of the system.

REFERENCES

1. Petrick, M., Recent developments in liquid-metal MHD power systems. *Power Reactor Technol.* **9**, 187–193 (1966).
2. Feldman, K. T., and Whiting, G. H., The heat pipe. *Mech. Engrg.* **89**, 30–33 (1967).
3. Busse, C. A., Optimization of heat pipe thermionic convecters for space power supplies. *Communaute Eur. Energ. At. EURATOM EUR (Rapp.)* **EUR2534**.e (1965).
4. Bussard, R. W., and DeLauer, R. D., "Nuclear Rocket Propulsion." McGraw-Hill, New York, 1958.
5. Keepin, G. R., "Physics of Nuclear Kinetics." Addison-Wesley, Reading, Massachusetts, 1965.
6. Mohler, R. R., and Shen, C. N., "Optimal Control of Nuclear Reactors." Academic Press, New York, 1970.
7. Pontryagin, L. S., Boltyanski, V. G., Gamkrelidze, R. V., and Mischenko, E. F., "The Mathematical Theory of Optimal Processes." Wiley (Interscience), New York, 1962.
8. Markus, L., Controllability of nonlinear processes. *SIAM J. Control* **3**, 78–90 (1965).
9. Eckert, E. R. G., and Drake, Jr., R. M., "Heat and Mass Transfer." McGraw-Hill, New York, 1959.
10. Humble, L. V., *et al.*, Measurements of average heat transfer and friction coefficients for subsonic flow of air in smooth tubes at high surface and fluid temperatures. *NACA Rept.* No. 1020. Nat. Adv. Comm. Aeronaut., Washington, D.C., 1951.
11. Ash, M., "Nuclear-Reactor Kinetics." McGraw-Hill, New York, 1965.
12. Murray, R. L., "Nuclear Reactor Physics," p. 180. Prentice-Hall, Englewood Cliffs, New Jersey, 1957.
13. Mohler, R. R., and Perry, Jr., J. E., Nuclear rocket engine control. *Nucleonics* **19**, 80–84 (1961).
14. Stepanoff, A. J., "Centrifugal and Axial Flow Pumps," 2nd ed. Wiley, New York, 1957.
15. Schultz, M. A., "Control of Nuclear Reactors and Power Plants." McGraw-Hill, New York, 1955.

16. Roberts, J. J., and Smith, Jr., H. P., Time-optimal solution to the reactivity-xenon shutdown problem. *Nuclear Sci. Engrg.* **22**, 470–478 (1965).
17. Rosztoczy, J. R., and Weaver, L. E., Optimum reactor shutdown program for minimum xenon buildup. *Nuclear Sci. Engrg.* **26**, 318 (1964).
18. Ash, M., Application of dynamic programming to optimal shutdown control. *Nuclear Sci. Engrg.* **24**, 77–78 (1966).

V

Ecologic and Physiologic Control

5.1 POPULATIONS OF SPECIES

The most simple population model has been introduced in Chapter I by Eq. (1.3), $dx/dt = ux$.† Then in Chapter IV neutron population generated by nuclear fission is described by Eqs. (4.2) to (4.7). Populations of species of organisms formed by reproduction, like neutron populations produced by nuclear fission, are controlled by a population coefficient, u = birth rate − death rate (in the case of nuclear fission called multiplication or reactivity). Naturally, population feedback around the bilinear process again makes the closed-loop system more highly nonlinear. Even for single-celled organisms, which reproduce by dividing, the birth rate may be manipulated.

A population of cells growing by cellular fission, just like a population of neutrons growing by nuclear fission, is apparently a stochastic process. And obviously the same is true throughout nature. Still the deterministic models discussed here are useful in that they describe average populations.

† Population models are discussed only briefly in order to introduce bilinear models in biology. Pielou [1] presents an excellent detailed analysis of population dynamics.

Population coefficients are manipulated in many ways to control populations of species. Resource and environmental changes may be made to affect birth rate or death rate directly. Certain control evolves in a natural manner and may be caused, for example, by limited resources in a competitive environment. Even in the absence of resource constraints, however, there may be naturally built-in inhibitors. Such a constraint was established by a controlled experiment on an island where deer were provided with all the food and water they wanted. Yet they only multiplied up to a certain population density. Such biological constraints are discussed by Markert [2].

Population constraints may be generated by population feedback around the bilinear model. A simple model may be derived as follows. Let ax be the rate of population growth with unlimited resources and no competition, and let a/b be the maximum attainable population. Then, assume that the actual rate of growth is the product of the potential rate multiplied by the unrealized proportion or

$$\frac{dx}{dt} = ax\left[\frac{(a/b) - x}{(a/b)}\right] = u(x)x, \tag{5.1}$$

where $u(x) = a - bx$ and x is population. Here a is called the intrinsic rate of natural growth, and (5.1) is called the Verhulst–Pearl logistic equation [1].

In nature, further inhibition may affect the parametric control by the presence of other competing species. The presence of a second species of population x_2 affects the population of the first species x_1 similarly, as is shown by

$$dx_1/dt = u_1(x_1, x_2)x_1, \tag{5.2}$$

where

$$u_1(x_1, x_2) = a_1 - b_{11}x_1 - b_{12}x_2,$$

and a_1, b_{11}, b_{12} are positive. Likewise the first species might constrain population of the second by

$$dx_2/dt = u_2(x_1, x_2)x_2, \tag{5.3}$$

where

$$u_2(x_1, x_2) = a_2 - b_{21}x_1 - b_{22}x_2,$$

and a_2, b_{21}, b_{22} are positive.

If one species is a parasite or predator and the other is its host or prey, their coupled population dynamics may be approximated by (5.2) and (5.3) with the slight alteration that b_{22} is negative since parasite population grows at the expense of its hosts. Also, in some cases it may be assumed that parasites do not compete with one another so that b_{22} is zero. On the other hand, a_2 will always be negative for the parasite and will be negative for the predator with no alternative food than the prey.

5.2 BIOCHEMICAL PROCESSES AND THE CELLULAR PLANT

Even at a more elementary level of nature, bilinear control is synthesized by catalysts in chemical reactions which are presumed to obey the law of mass action, that is, the reaction rate is proportional to concentrations of the reacting species with the catalytic agent entering the dynamics as a multiplicative control similar to the population coefficient. While these reactions generally are reversible with different rate constants in the two directions, the catalytic control mechanism may greatly increase the forward reaction. (An increase by several orders of magnitude is not uncommon.) The reaction is said to be in the steady state when the forward reaction balances the backward reaction. Therefore, the catalyst must be introduced in a carefully controlled manner if the reaction is to proceed rapidly and then attain some equilibrium.

Biochemical reactions such as those that are generated in the human body in conjunction with metabolism or synthesis of chemical species are controlled by an enzyme catalyst. Without the enzyme the system could not attain the necessary degree of controllability or sustain life. Again, the need for multiplicative control is obvious.

Even an individual cell may have thousands of series and parallel reactions taking place internally. Many of the reactions take place so rapidly, however, that their reaction times can be neglected and only a small number of pacemakers determine speed of response. Within the cell itself, component processes under the control of many enzymes seem to proceed in a most efficient manner. There is no overproduction but rather a series of readjustments [3].

The Michaelis–Menton equations are commonly used to approximate such biochemical reactions. Following directly from the law of mass action they are given by

$$dx_1/dt = a_1x_2 - bux_1 \tag{5.4}$$

and

$$dx_2/dt = bux_1 - (a_1 + a_2)x_2, \tag{5.5}$$

where x_1, x_2 are concentrations of a chemical substrate (substance acted upon by an enzyme) and a complex, u is enzyme concentration, and a_1, a_2, b are positive rate constants. Again, the model is bilinear, and similar to population dynamics, various feedback controllers have been hypothesized and partly substantiated by experiments. The common Michaelis–Menton model assumes that the complex formed by combination of enzyme and substrate decomposes into an enzyme and end product of concentration x_3 so that

$$dx_3/dt = a_2x_2 \tag{5.6}$$

and

$$du/dt = (a_1 + a_2)x_2 - bux_1 \tag{5.7}$$

with $x_1 + x_2 + x_3 = s_0$ and $u + x_2 = e_0$. Here, s_0 and e_0 are positive constants arising from the total substrate and total enzyme available in the cell.

Enzyme control can be effected by manipulation of either total amount of enzyme present or amount present at a given time. The latter may be generated by means of competitive inhibition with a second substrate or the end product competing for the supply of enzyme. The total amount of enzyme in the cell is increased (induction) or decreased (repression) through the DNA. Again, there are two classes of enzymes for control, namely, constitutive enzymes and inducible enzymes. The former are directly generated in controlled quantity by the DNA along with details of enzyme structure. The second class is synthesized in controlled fashion by the delayed action of some inducer molecule or substrate. It is the feedback through the DNA which results in the primary adaptive or variable-structure control process. More details of this system-identification problem are given by Eigen and Hammes [4].

5.3 COMPARTMENTAL MODELS

A single cell frequently is a convenient compartment for the purpose of mathematical modeling of a more complicated organism. Nutrients are generated for cell growth and cell division by biochemical processes within the cell itself and by transfer across the cell membrane. The cell stores some starch and fat to be consumed in the absence of external stores. But as shown by the cellular dynamics in the previous section, enzymes must be present to activate the necessary reactions which lead to the formation of proteins and nucleic acids and their precursors. These in turn provide cell growth. Water may be generated by metabolism within the cell itself or may be transferred across the membrane from outside the cell.

Molecules of compounds formed from the biochemical processes within the cell move in all directions in a random manner. Molecules diffuse naturally from a region of high concentration to one of lower concentration. Consequently, within a single cell and within a complex of cells the balance between chemical reaction and natural diffusion leads to a highly concentrated core with a steady outward flow.

This transfer of so-called metabolites from one compartment to another across some membrane takes place at a rate nearly proportional to the difference concentrations in the two compartments if membrane permeability is identical in both directions [5]. Besides the bilinear control by enzymes on the biochemical reactions within cells, parametric or bilinear control is synthesized by the control action of enzymes that cause changes in membrane permeability.

5.3.1 *Model Description*

Suppose there are two compartments of capacities c_1 and c_2, uniform concentrations $\xi_1(t)$ and $\xi_2(t)$, and amounts $x_1(t)$ and $x_2(t)$ of substance X. Then the net flux diffusing through the membrane from compartment 2 to compartment 1 is described by

$$\phi_{12} = \rho_{12}\xi_1(t) - \rho_{21}\xi_2(t), \tag{5.8}$$

or

$$\phi_{12} = (\rho_{12}/c_1)x_1(t) - (\rho_{21}/c_2)x_2(t) \tag{5.9}$$

and

$$\phi_{12} = -\phi_{21}, \tag{5.10}$$

where ρ_{12} and ρ_{21} are membrane exchange coefficients (not generally of the same value). It is assumed that the membrane has no storage of substances.

In addition to the natural diffusion mechanism for material transport from regions of high concentration to regions of low concentration, experimental evidence shows that sometimes there exists an active transport from low concentration to high concentration or a retention of high concentration of various substances by cells against appreciable gradients. The necessary energy for this active transport is received from the cellular metabolism. Hill and Kedem [6] model the net active transport of a single substance across a membrane by

$$\frac{a(t)\xi_2(t) - b(t)\xi_1(t)}{c + d\xi_1(t) + e\xi_2(t)}, \tag{5.11}$$

where it is assumed that a, b may be manipulated, and c, d, e are constants. While active transport is obviously nonbilinear it may be approximated by the linear terms of a Taylor series so that the overall process is approximated by a system of bilinear equations with $a(t)$, $b(t)$ as control variables and $\xi_1(t)$, $\xi_2(t)$ as state variables. Similarly, material transport due to hydrostatic pressure differences may be closely approximated by a bilinear model.

Suppose a single substance X is distributed among n compartments (cells, spaces, phases, pools). The rate of change of substance amount in the ith compartment is described by the conservation equations:

$$dx_i/dt = \sum_{l=1}^{n}{}' \phi_{il}(t) - \sum_{q=1}^{n}{}' \phi_{qi}(t) + \phi_{ia}(t) - \phi_{ai}(t) + p_i(t) - d_i(t), \qquad i = 1, \ldots, n, \tag{5.12}$$

where the primed summation sign indicates that the ith term of the summation is deleted, $x_i(t)$ is the amount of substance X in compartment i, ϕ_{ij} is the flux of X from compartment j to compartment i, $\phi_{ia}(t)$ [$\phi_{ai}(t)$] is the flux of X into [out of] compartment i from [to] the environment, $p_i(t)$ [$d_i(t)$] is the rate of production [destruction] of X in compartment i.

Equation (5.12) may be put into a more compact form so that environmental fluxes, production terms, and destruction terms are combined with compartmental fluxes. In this manner,

$$dx_i/dt = \sum_{l=1}^{n+2}{}' \phi_{il}(t) - \sum_{q=1}^{n+2}{}' \phi_{qi}(t), \qquad i = 1, \ldots, n, \tag{5.13}$$

where $\phi_{i,n+1}(t) = \phi_{ia}(t)$, $\phi_{n+1,i}(t) = \phi_{ai}(t)$, $\phi_{i,n+2}(t) = p_i(t)$, and $\phi_{n+2,i}(t) = d_i(t)$.

If it is assumed that compartmental capacities do not change significantly over the time period of interest [see (5.9)], the flux terms can be expressed in terms of time-invariant bilinear equations of the general form of

$$\phi_{il}(t) = \sum_{j=1}^{n} a_{il,j}x_j(t) + \sum_{j=1}^{n}\sum_{k=1}^{m} b_{il,jk}x_j(t)u_k(t)$$
$$+ \sum_{k=1}^{m} c_{il,k}u_k(t) + v_{il}(t), \tag{5.14}$$

for $i, l = 1, \ldots, n+2, i \neq l; (i, l) \neq (n+1, n+2)$ or $(n+2, n+1)$. For some problems $v_{il}(t)$ may be neglected or may be a given time function. In other cases, production and destruction terms may be additive controls. To distinguish this system as a slightly different class of bilinear system, let the general bilinear model be described by

$$d\mathbf{x}/dt = \mathbf{A}\mathbf{x} + \sum_{k=1}^{m} \mathbf{B}_k u_k \mathbf{x} + \mathbf{C}\mathbf{u} + \mathbf{v}, \tag{5.15}$$

where $\mathbf{v}$ is an n-vector and the other terms were described previously. Equations (5.13) and (5.14) are put into this general bilinear form if $\mathbf{A}$ has elements

$$a_{ij} = \sum_{l=1}^{n+2}{}' a_{il,j} - \sum_{q=1}^{n+2}{}' a_{qi,j},$$

$$b_{i,jk} = \sum_{l=1}^{n+2}{}' b_{il,jk} - \sum_{q=1}^{n+2}{}' b_{qi,jk}, \qquad k = 1, \ldots, m,$$

$$c_{ik} = \sum_{l=1}^{n+2}{}' c_{il,k} - \sum_{q=1}^{n+2}{}' c_{qi,k}$$

$$v_i = \sum_{l=1}^{n+2}{}' v_{il}(t) = \sum_{q=1}^{n+2}{}' v_{qi}(t).$$

While real processes may involve transport of more than one substance, no loss of generality is obtained in regards to the bilinear form of the system equations above.

The concept of compartmentation has been applied for anatomical and mathematical convenience here. Still this does not imply that optimal discretization of the distributed system in a mathematical sense would yield the same compartments.

In the analysis above, it is implicitly assumed that the process involves mass transport. Again, the concept is more general. Similar compartmental models have been used to study such diverse processes as body fluid and electrolyte balance [7], thermal regulation [8], kinetics of material injected or ingested into or excreted from the body [9], and control of carbon dioxide in the lungs [10], as well as kinetics of metabolites in cell suspensions or tissues [11].

Homeostasis, as defined by Cannon [12], is the process of regulation in organisms to maintain an essentially constant internal environment. For the compartmental models discussed above, homeostasis is the regulation of substance amounts or concentrations by means of adjusting membrane permeabilities and/or rates of production and destruction. This regulatory mechanism is studied in conjunction with regulation of temperature and carbon dioxide later in this chapter. Indeed, the major portion of this chapter is devoted to what might be called homeostasis in compartmental bilinear physiological systems.

By their nature, these systems are difficult to isolate for purposes of direct identification and modeling. It is frequently impossible to observe state or control variables, yet manipulate control variables as is so often desired for system identification. The compartmented structure of these systems, however, does lend itself conveniently to tracer analysis which in turn can be used for system identification.

5.3.2 *Tracer Experiments*

A tracer substance in effect puts a tag on some quantity for which it is desired to study dynamic behavior. The method is not new, for sheepherders have long tagged certain sheep with bells in order to trace the wanderings of their flocks. Similarly, radar is utilized to track and identify targets.

Isotopic tracers are particularly convenient and usually harmless to identify compartment capacities and substance amounts and concentrations. Usually it is assumed (1) that the tracer is distributed uniformly throughout the substance X, (2) that labeled and unlabeled X behave identically, and (3) that the tracer does not alter the system dynamics. Obviously, this requires that the amount of tracer in the system is neglected.

The portion of tracer in the total amount of X in compartment i, called the specific activity a_i, is described by [13, 14]

$$d(x_i a_i)/dt = \sum_{l=1}^{n}{}' \phi_{il}(t)a_l - \sum_{q=1}^{n}{}' \phi_{qi}(t)a_i + \phi_{ia}(t)a_{ia}(t) - \phi_{ai}(t)a_i - d_i(t)a_i + f_i(t), \qquad i = 1, \ldots, n, \quad (5.16)$$

where $a_{ia}(t)$ is the specific activity of the substance in influx $\phi_{ia}(t)$, and $f_i(t)$ is the influx of tracer inserted directly into compartment i. If $\phi_{ia}(t) \equiv 0$, then $a_{ia}(t) \equiv 0$. It is interesting and convenient that the tracer system (5.16) is linear in specific activity even if the physiological model is nonlinear.

Ordinarily tracer experiments are conducted when the compartmental system is in steady state with $\dot{x}_i = \dot{\phi}_{ij} = \dot{\phi}_{ai} = \dot{\phi}_{ia} = \dot{p}_i = \dot{d}_i = 0$, $i, j = 1, \ldots, n$, $i \neq j$.† Also, it is often assumed that the system is closed with $\phi_{ai} = \phi_{ia} = 0$, and conservative with $p_i = d_i = 0$, $i = 1, \ldots, n$. Therefore, the unforced tracer dynamics are time invariant, linear, and take the form of

$$d\mathbf{a}/dt = \mathbf{S}\mathbf{a}, \S \qquad (5.17)$$

where $\mathbf{a}$ is a column vector of components $a_1, \ldots, a_n$, and $\mathbf{S}$ is an $n \times n$ matrix of elements

$$s_{ij} = \phi_{ij}/x_i \geq 0, \qquad i, j = 1, \ldots, n, i \neq j,$$
$$s_{ii} = -(1/x_i) \sum_{l=1}^{n}{}' \phi_{il} = -(1/x_i) \sum_{q=1}^{n}{}' \phi_{qi}, \qquad i = 1, \ldots, n. \quad (5.18)$$

† This condition is not referred to as equilibrium in the physiology literature since there still may be net flux between compartments and with the environment.

§ See (5.27)–(5.31) for the general form of tracer dynamics, from which (5.17), (5.23), and (5.25) may be taken directly.

Specification of necessary data to identify fluxes from tracer observations and on the solution of the fluxes with complete [13, 14] and incomplete [15] data has been derived.

5.3.3 *Identification by Tracers*

For most physiological processes it is not at all convenient to decouple subsystems, to measure all controls, and to measure all state variables. Usually, it is not practical to add test signals to the physiological process that is to be modeled. Unfortunately, this precludes the use of many of the most convenient properties of bilinear systems. Such limitations are especially prevalent in human-body experiments. For many of these processes, compartmentation lends itself conveniently to modeling and identification by means of tracer experiments since such experiments may be conducted without having a noticeable effect on the regulated process. Added to this is the convenience of linearity exhibited by the tracer kinetics.

Even the most basic identification problem, that of estimating system order, can be quite complicated for an inaccessible physiological system. Even with tracers, it should be ascertained that there is adequate experimental access before data are extracted for identification purposes.

There is a common theorem in linear system theory which states that the order n of a linear system can be obtained from input–output observations if and only if the system is completely controllable (CC) and completely observable (CO) [16]. For convenience, let the linear system be described by

$$d\mathbf{x}/dt = \mathbf{Ax} + \mathbf{Cu} \tag{5.19}$$

and

$$\mathbf{y} = \mathbf{Dx} + \mathbf{Fu}, \tag{5.20}$$

where again $\mathbf{x}$ is the n-dimensional state vector, $\mathbf{u}$ is the m-dimensional control vector, $\mathbf{y}$ is the q-dimensional output vector, and matrices $\mathbf{A}$, $\mathbf{C}$, $\mathbf{D}$, and $\mathbf{F}$ are of the appropriate dimensions.

Controllability of this system is analyzed in Chapter II. As noted there the linear system is CC if and only if the rank of the $n \times (m \cdot n)$ matrix

$$\mathbf{E} = [\mathbf{C} \mid \mathbf{AC} \mid \cdots \mid \mathbf{A}^{n-1}\mathbf{C}] \tag{5.21}$$

is n. A similar test derived for observability states that the linear system of Eqs. (5.9) and (5.10) is CO if and only if the $n \times (q \cdot n)$ matrix

$$\mathbf{G} = [\mathbf{D}^{\mathrm{T}} \mid \mathbf{A}^{\mathrm{T}}\mathbf{D}^{\mathrm{T}} \mid \cdots \mid \mathbf{A}^{\mathrm{T}n-1}\mathbf{D}^{\mathrm{T}}] \tag{5.22}$$

has rank n [16].

Another way of stating the desired theorem for application here is that the lowest order of the linear system that will "realize" the observed input–output data is $n = n_o$ if and only if both **E** and **G** have rank n. Otherwise, $n > n_0$. Such lowest order systems are called minimal realizations. While a system based on input–output data may generally be identified by any number of realizations, we are only concerned with minimal realizations in this book.

Now it is desired to apply this result to the tracer experiment at hand. Tracer can be inserted by special routes, such as by injection, directly into a particular compartment, or tracer can be ingested by normally existing pathways.

First, suppose tracer is inserted directly into p system compartments and that its behavior can be monitored in q compartments. Define an input matrix **P** with unity elements in each row corresponding to a compartment accessible for tracer insertion and with zero elements elsewhere. Each column of **P** has exactly one unity element and the rank of **P** is p. Also, define a $q \times n$ output matrix **Q** with zero elements except for unity in each column which corresponds to an observable compartment. Then, from (5.16) or (5.17) experimental access is described by

$$d\mathbf{a}/dt = \mathbf{S}\mathbf{a} + \mathbf{X}_{\mathrm{d}}^{-1}\mathbf{P}\mathbf{f}_{\mathrm{i}} \tag{5.23}$$

and

$$\mathbf{w} = \mathbf{Q}\mathbf{a}, \tag{5.24}$$

where $\mathbf{X}_{\mathrm{d}}$ is the diagonal matrix of elements $x_1, \ldots, x_n$ (assumed to be at steady state); $\mathbf{f}_{\mathrm{i}}$ is the p-dimensional vector of inserted total activity fluxes, and therefore $\mathbf{X}_{\mathrm{d}}^{-1}\mathbf{P}\mathbf{f}_{\mathrm{i}}$ is the vector of inserted specific activity fluxes; and $\mathbf{w}$ is the q-dimensional output vector of observed compartmental specific activities.

From (5.21) and (5.22), it is apparent that (5.23) and (5.24) are CC and CO, respectively, if and only if

$$[\mathbf{P} \mid \mathbf{S}\mathbf{P} \mid \cdots \mid \mathbf{S}^{n-1}\mathbf{P}]$$

and

$$[\mathbf{Q}^{\mathrm{T}} \mid \mathbf{S}^{\mathrm{T}}\mathbf{Q}^{\mathrm{T}} \mid \cdots \mid \mathbf{S}^{\mathrm{T}n-1}\mathbf{Q}^{\mathrm{T}}]$$

are of rank n, respectively. Examples showing application of the rank test to several compartmental systems are given in Fig. 5.1. For case 4, the conservative system, it is assumed that the fluxes divided by the amount of substance from the originating compartment are all equal (i.e., $\phi_{12}/x_1 = \phi_{21}/x_2 = \cdots$).

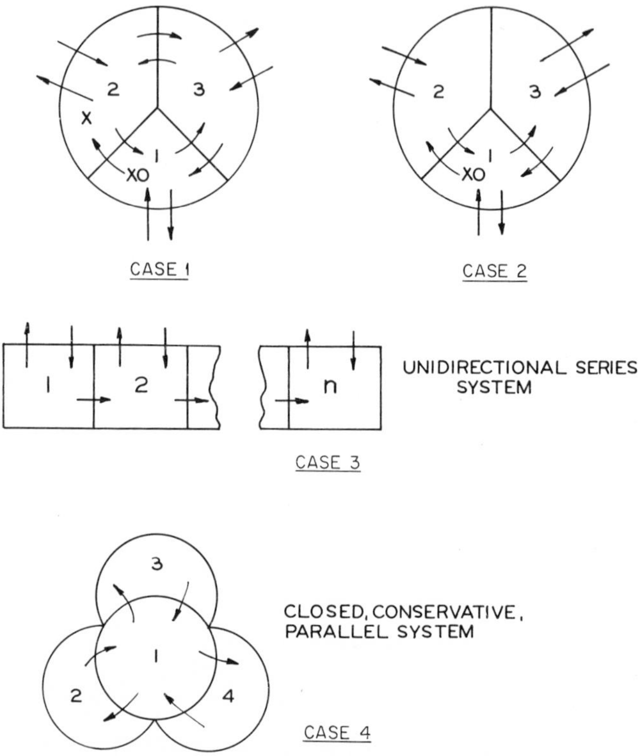

Fig. 5.1. Tracer tests of controllability and observability: → denotes flux; x denotes tracer insertion; o denotes tracer observation. *Case 1*: Not complete controllability (CC), not complete observability (CO). *Case 2*: Not CC, not CO, but CC, CO with either 2 or 3 (instead of 1) accessible. *Case 3*: CC if and only if tracer inserted in 1, CO if and only if tracer observed in *n*. *Case 4*: Need access to at least two compartments for CC and CO.

Now consider the experiment with tracer entering the system by naturally available flux routes. Assume that fluxes ϕ_{ia} with specific activities a_{ia}, $i = 1, \ldots, n$, originate from p' separate external sources of labeled substance, each with its own tracer specific activity a_i', $i = 1, \ldots, p'$. In this case, (5.17) becomes

$$d\mathbf{a}/dt = \mathbf{Sa} + \mathbf{X}_{\mathrm{d}}^{-1}\mathbf{\Phi}_{ad}\mathbf{P}'\mathbf{a}', \tag{5.25}$$

and again

$$\mathbf{w} = \mathbf{Qa} \tag{5.26}$$

where $\mathbf{\Phi}_{ad}$ is a diagonal matrix of elements $\phi_{1a} \cdots \phi_{na}$, $\mathbf{a}'$ is a p'-dimensional tracer source activity vector, and $\mathbf{P}'$ is the $n \times p'$ output matrix with a single unity in each row i for which $a_{ia} \neq 0$. Since more than one compartment may receive substance from the same external source, $\mathbf{P}'$ may have more than a single unity in a given column. The rank test for complete observability is as before, but the $\mathbf{E}$ matrix for determination of complete controllability is more complicated with (5.25) CC if and only if

$$[\mathbf{X}_{\mathrm{d}}^{-1}\mathbf{\Phi}_{ad}\mathbf{P}' \mid \mathbf{SX}_{\mathrm{d}}^{-1}\mathbf{\Phi}_{ad}\mathbf{P}' \mid \cdots \mid \mathbf{S}^{n-1}\mathbf{X}_{\mathrm{d}}^{-1}\mathbf{\Phi}_{ad}\mathbf{P}']$$

has rank n. If no two compartments receive tracer from the same external source, then this matrix is the same as for the previous case with $\mathbf{P}'$ replacing $\mathbf{P}$.

Commonly used tracer "soak-up" and "washout" experiments fit the latter category of "natural" tracer absorption. The system might be placed in a radioactive bath, and its absorption of tracer monitored in a "soak-up" experiment. In tracer "washout" experiments, the system is saturated to a given specific activity of tracer, and then the system is put in a tracer-free environment so that washout is observed. In either case, the presence of all compartments cannot be detected unless the rank test is satisfied.

The value of n may be estimated from a measurement of the state transition matrix if this is available. Or more commonly, after showing that the tracer system is CC–CO, an impulse response matrix or transfer function matrix may be appropriately fitted to ascertain eigenvalues and system order.

In the physiological literature, it is commonly assumed that all system eigenvalues are real in arriving at exponential fits to the response. It is very easy to show, however, that such an assumption can lead to very erroneous conclusions. From a theorem by Gersgorin (see Lancaster [17, p. 226]) it can be shown that the eigenvalues of the tracer matrix **S** are in the left half of the complex plane or at the origin. Detailed properties of tracer dynamics are presented by Hearon [18] and by Smith [19], and the discussion here follows the latter work.

When the compartmental system is not in steady state, the general tracer system for direct insertion and "natural" absorption takes the form of

$$d\mathbf{a}/dt = \mathbf{S}(t)\mathbf{a} + \mathbf{R}(t)\mathbf{v} \tag{5.27}$$

and

$$\mathbf{w} = \mathbf{Q}\mathbf{a}, \tag{5.28}$$

where in general $\mathbf{S}(t)$ has elements

$$s_{ij}(t) = \phi_{ij}(t)/x_i(t) \geq 0, \qquad i, j = 1, \ldots, n, \qquad i \neq j, \tag{5.29}$$

$$\begin{aligned} s_{ii}(t) &= -\frac{1}{x_i(t)}\left(\sum_{l=1}^{n}{}' \phi_{il}(t) + \phi_{ia}(t) + p_i(t)\right), \\ &= -\frac{1}{x_i(t)}\left(\sum_{q=1}^{n}{}' \phi_{qi}(t) + \phi_{ai}(t) + d_i(t) + \dot{x}_i(t)\right) \end{aligned} \tag{5.30}$$

$$i = 1, \ldots, n.$$

$\mathbf{R}(t)$ is an $n \times (p + p')$ matrix which is defined by

$$\mathbf{R}(t) = [\mathbf{X}_{\mathrm{d}}^{-1}(t)\mathbf{P}\mathbf{X}_{\mathrm{d}}^{-1}(t)\mathbf{\Phi}_{a\mathrm{d}}(t)\mathbf{P}'], \tag{5.31}$$

$\mathbf{v}$ is a $(p + p')$ input vector, and the remaining terms are as previously defined. Here $\mathbf{v}$ is composed of previously defined f_i and $\mathbf{a}'$.

Depending on the form of matrices **P** and **P′**, some of the state variables $x_i(t)$ and some of the influxes $\phi_{ia}(t)$, $i = 1, \ldots, n$, may be determined from $\mathbf{R}(t)$. Most of the identification information, however, is obtained from $\mathbf{S}(t)$, the tracer system matrix. Common methods of linear system realization and parameter estimation by error minimization may be used to identify $\mathbf{S}(t)$. Only an introduction to this difficult identification problem is presented here. No consideration is given to the presence of noise, which greatly complicates the problem. Indeed, certain system parameters may be computed algebraically from the tracer equations and the compartmental state equations if sufficient information is available.

When it is assumed that the compartmental system is closed, conservative, and in steady state (and this is a frequent assumption), **S** is a constant matrix defined as for (5.18). In principle at least, the realization problem is solved algebraically from well-known solutions to a time-invariant linear system [16].

Once the elements of matrix $\mathbf{S}(t)$ for a non-steady-state condition are available on some interval $I = [t_0, t_f]$, the system state and compartmental flux parameters may sometimes be computed from (5.29) and (5.30). If the production rates $p_i(t)$ and influxes $\phi_{ia}(t)$, $i = 1, \ldots, n$, are known on I, these equations show that the state can be obtained from

$$x_i(t) = -\frac{\phi_{ia}(t) + p_i(t)}{\sum_{l=1}^{n} s_{il}(t)}, \qquad i = 1, \ldots, n. \tag{5.32}$$

Then, intercompartmental fluxes may be found on I by

$$\phi_{ij}(t) = s_{ij}(t)x_i(t), \qquad i, j = 1, \ldots, n, \qquad i \neq j. \tag{5.33}$$

The conservative compartmental system, of which the water balance system studied later is a good example, is described by

$$\frac{dx_i}{dt} = \sum_{l=1}^{n}{}' \phi_{il}(t) - \sum_{q=1}^{n}{}' \phi_{qi}(t), \qquad i = 1, \ldots, n, \tag{5.34}$$

if it is closed. While closed living systems do not really occur, physiological systems frequently are placed in a closed environment for observation. Then, an $(n-1)$th-order open physiological system, in conjunction with its environment, becomes an nth-order closed compartmental system.

The tracer dynamics corresponding to (5.34) are defined by

$$d\mathbf{a}/dt = \mathbf{S}(t)\mathbf{a}, \tag{5.35}$$

where $\mathbf{S}(t)$ has elements

$$\begin{aligned} s_{ij}(t) &= \phi_{ij}(t)/x_i(t) \\ s_{ii}(t) &= -\frac{1}{x_i(t)} \sum_{l=1}^{n}{}' \phi_{il}(t) \\ &= -\frac{1}{x_i(t)} \left(\sum_{q=1}^{n}{}' \phi_{qi}(t) + \dot{x}_i(t) \right), \end{aligned} \tag{5.36}$$

$$i, j = 1, \ldots, n.$$

Then, from $\mathbf{S}(t)$, the state may be ascertained on $I = [t_0, t_f]$ from the adjoint equation,

$$\dot{\mathbf{x}} = d\mathbf{x}/dt = -\mathbf{S}^{\mathrm{T}}(t)\mathbf{x}, \tag{5.37}$$

where $\mathbf{x}(t_1) = \mathbf{x}_1$, $t_1 \in I$. If the compartmental system is at steady state, (5.37) becomes an algebraic equation,

$$\mathbf{S}^{\mathrm{T}}\mathbf{x} = \mathbf{0}, \tag{5.38}$$

where $\mathbf{S}$ is a singular matrix since

$$\sum_{j=1}^{n} s_{ij} = 0, \qquad i = 1, \ldots, n.$$

Hence, (5.38) has a nontrivial solution for $\mathbf{x}$.

For the time-variant case, even the test for a minimal realization is greatly complicated. Following D'Angelo [20], convenient test matrices associated with (5.27) are

$$[\mathbf{R}(t) \mid \mathbf{\Delta}_c\mathbf{R}(t) \mid \cdots \mid \Delta_c^{n-1}\mathbf{R}(t)], \tag{5.39}$$

where

$$\mathbf{\Delta}_c = -\mathbf{S}(t) + d/dt,$$

and

$$[\mathbf{Q}^{\mathrm{T}} \mid \mathbf{\Delta}_0\mathbf{Q}^{\mathrm{T}} \mid \cdots \mid \mathbf{\Delta}_0^{n-1}\mathbf{Q}^{\mathrm{T}}], \tag{5.40}$$

where

$$\Delta_0 = \mathbf{S}^{\mathrm{T}}(t) + d/dt,$$

and it is implicitly assumed that $\mathbf{S}(t)$ and $\mathbf{R}(t)$ are differentiable $n - 2$ and $n - 1$ times. Then, it is shown by D'Angelo [20, p. 134] that time-variant system [5.27] is a minimal realization on $I = [t_0, t_f]$ if the rank of the matrices defined by (5.39) and (5.40) is n on I. These conditions are sufficient but not necessary.

Algorithms are available to synthesize minimal realizations of time-variant systems from input–output observations. D'Angelo [20, p. 183] gives methods for generating state, input, and output matrices from the impulse response matrix for a time-variant linear system, and Chen [16] provides methods which provide similar matrices from the transfer function matrix for a time-invariant linear system. These synthesis procedures specify exactly what experiments need to be conducted to calculate the desired realizations. Consequently, the methods are quite systematic. The time-variant realization, however, requires the impulse response matrix $H(t, \tau)$ which may be difficult to procure experimentally. Then $H(t, \tau)$, which may be quite noisy in practice, must be multiply differentiated with respect to both variables.

In addition to the direct algebraic and synthesis methods, the error minimization method discussed briefly in Chapter I may be utilized. The ensuing optimization problem can be solved by iterative computer algorithms such as quasi-linearization.

5.4 WATER BALANCE

Water, on the average comprising about 60 percent of the body weight, is the principal constituent of the human body. Water content varies widely among individuals (between 45 to 75 percent of body weight) since it depends on the quantity of fatty tissue, but it is very nearly constant for a given person.

About two-thirds of the body water is distributed throughout the body within the cells, while the other one-third is located in the so-called extracellular compartment. The extracellular compartment includes the plasma or intravascular compartment, or the interstitial compartment (in which tissue cells are bathed) and the transcellular compartment. There is a distinct anatomical boundary, such as wall membrane or blood vessel wall, for some compartments, but for the interstitial compartment, anatomical boundaries are not so clear. The transcellular compartment, which includes digestive secretions, is separated from plasma by a continuous layer of epithelial cells as well as by the capillary membrane.

Distribution of body water throughout these compartments is obtained by tracers which are injected into the compartment. Some compartments, such as the intracellular one, cannot be so isolated and must be calculated from differences. Deuterium oxide or tritiated water injected intravenously is used commonly to estimate the total amount of body water. Concentration of deuterium oxide is estimated by specific gravity measurement, and that of tritiated water by liquid scintillation counting. Plasma water is measured by radioiodinated human serum albumin (RISA) or by means of a blue dye (T-1824) which binds tightly to the subject's plasma albumin. While extracellular water is difficult to ascertain, mannitol is probably the most used tracer here. Then, intracellular water is calculated as the difference between the total and the extracellular water. Also, there is a small quantity of stored water not readily exchangeable because it is bound in the formation of such compounds as estrogen and progesterone and the constituents of connective tissue.

Again, water balance throughout the entire system is regulated by homeostasis through pressure forces and membrane permeabilities. Output feedback to the control is established in a somewhat complicated manner that involves the hypothalamus portion of the brain. To get an idea of the response time of this control mechanism internal to man, note that an amount of water equal to two-thirds of the blood volume is exchanged between the blood and the extravascular fluid each minute, while labeled water in tracer experiments is entirely equilibrated with total body water in a few hours [21, p. 627].

Water is taken into the body's extracellular compartment by drinking, by eating, by skin absorption, and by respiratory absorption. Oxidative water, a metabolic end product which varies with diet and metabolic rate, is an influx to the intracellular compartment. Effluxes from the extracellular compartment to the environment mainly include urine, fecal water, skin evaporation, and respiration.

It is apparent that man, along with the animal world, has instinctive control over these influxes and to some lesser extent over the effluxes. Besides the distribution of water between compartments, homeostasis acts through the hypothalamus to control total body water influx and efflux. The kidney with its thousands of complex filtration passages called nephrons is a key component in the water balance control mechanism.† For example, if the body takes in an excess of water, the kidney compensates by producing a large amount of dilute urine. If water intake is deficient, a minimum amount of high-concentration urine is excreted. Of course, there are constraints on the renal capacity, and large deficits in water balance must be compensated for by thirst and increased intake. And, a minimum quantity of urine must be excreted to rid the body of nitrogenous wastes.

As an indication of how coupled physiological processes may be and

† See Pitts [22] or Smith [23] for a detailed description of kidney physiology.

how useful the kidney is as a control device, it might be noted here that the kidney also acts to regulate electrolyte balance, acid balance, waste excretion, hormone synthesis, and detoxication.

For water balance, however, the main function of the kidney is to process blood plasma through its nephrons to form urine. Within each kidney nephron, water loss is regulated by the action of antidiuretic hormone (ADH) on tubule and collecting duct permeability. Also glomerular† filtration rate (GFR) and renal plasma flow (RPF) may be regulated to control water loss. Renal plasma flow depends on arteriolar resistance in the kidney which is controlled by a network of constrictor muscles to control the process by parametric or approximately bilinear control similar to that used for temperature regulation in Section 5.6. Glomerular filtration rate, GFR, like RPF is a nonlinear function of arterial pressure which may be approximated by a linear function for typical deviations about the controlled level.

The net urinary water efflux ϕ_u is the difference between water filtration rate into the tubules at the glomerulus and the rate of water reabsorption (WRR) by the renal tubules, or

$$\phi_u = (\text{GFR}) - (\text{WRR}), \tag{5.41}$$

where

$$\text{GFR} = \rho_g \Delta p_g; \tag{5.42}$$

ρ_g is glomerulus membrane permeability. From the Starling hypothesis [20, p. 625] Δp_g is the net filtration pressure between the glomerular capillary and tubule hydrostatic pressures, $p_{gc} - p_t$, less the difference between the corresponding osmotic pressure, $\pi_{gc} - \pi_t$. Again, permeability is a bilinear control variable if the pressures are taken to be state variables or linear functions of state, with

$$\text{GFR} = \rho_g(p_{gc} - p_t) - (\pi_{gc} - \pi_t). \tag{5.43}$$

p_{gc} is normally the dominant term in (5.43) so that

$$\text{GFR} \approx \rho_g p_{gc}. \tag{5.44}$$

† The glomerulus is a renal capillary nodule located at the arteriole end of the nephron to collect an ultrafiltrate of the blood plasma.

Following Smith [19], p_{gc} is some fraction of kidney arterial pressure with the fraction determined by vasomotor-controlled arteriolar resistance of the kidney circulation. The kidney arterial pressure, in turn, is roughly a linear function of the amount of body water w_b. Consequently, glomerular filtration rate may be approximated by

$$\text{GFR} = u_g(c_g + b_g w_b), \tag{5.45}$$

where parametric control u_g is a product of permeability and arteriolar resistance terms; and c_g and b_g are constants with b_g positive. The water reabsorption term in (5.41), WRR, may be approximated as a controllable fraction of GFR so that net urinary efflux may be approximated by

$$\phi_u = (u_g + u_t)(c_g + b_g w_b), \tag{5.46}$$

where

$$u_t = k_t \rho_t \pi_k u_g, \tag{5.47}$$

and ρ_t is an effective tubule permeability governed by the presence of ADH and possibly other hormones, π_k is an average osmotic pressure driving water from the renal tubules, and k_t is a constant.

In summary, water excretion in urine may be approximated by a bilinear system of the form

$$\phi_u = u_k(c_g + b_g w_g), \tag{5.48}$$

where u_k is a sum of complicated control processes, and body water w_b may be a sum of compartmental amounts of water which in turn may be state variables. Physiologically, the control $u_k(t)$ involves a product of more basic control variables.

Water balance, to a much lesser degree, is also regulated through the skin and through the lungs. The mechanisms are modeled by Smith [19]. Again, the net water losses are regulated by parametric or multiplicative control functions. For the skin, net water losses are controlled by vasomotor control of blood flow and by diffusion permeability of the skin. Net loss in the lungs may be controlled parametrically by manipulation of ventilation rate and by vasomotor control of respiratory blood vessels.

5.5 TEMPERATURE REGULATION

5.5.1 Introduction

Man, as well as any other warm-blooded animals (including mammals and birds), utilizes homeostasis to maintain constant temperature in an ever-changing environment. To achieve such regulation, internal control mechanisms vary both heat losses and heat generation. For example, a decrease in ambient temperature results in an increase in body metabolism to generate more heat. The opposite effect is seen for cold-blooded animals, for whom internal temperature tends to follow ambient temperature. Temperature is regulated only passively in cold-blooded creatures, and consequently metabolism decreases with temperature.

Temperature dependence has been neglected for all the previous processes analyzed in this chapter.

Indeed, one of the strongest reasons for thermoregulation is to maintain a constant temperature for efficient cellular metabolism and other biochemical processes.

While temperature is distributed more or less continuously throughout the body, this distributed process, which gives rise to a system of partial differential equations, may be broken down into a finite number of lumps or compartments. It was shown for the heat exchange processes studied in Chapter IV that energy balances may be made on each compartment to derive the dynamical equations which define the state. In this case the fluxes considered for the compartmental systems in Section 5.3 are heat fluxes.

Apparently, it is most important that body core temperature is kept nearly constant. Unfortunately, very little basic knowledge is available on the need for a precise constant temperature such as that normally maintained in the body core. Neither is the reason for fever during disease understood. It seems reasonable that the thermoregulator should receive a new reference temperature in order that body processes function efficiently with respect to a new performance index which involves the disease-fighting capability of the body. On the other hand, it is not too unreasonable to expect the disease to cause a deterioration of certain control parameters so that efficient homeostasis is somewhat sacrificed. The latter argument, of course, is more consistent with our policy of relieving the fever by aspirin, and/or cold baths. In any event, control is again most important.

Body temperature regulation and its corresponding physiology are discussed in detail by Hardy [24] and Milsum [8]. Only a cursory analysis of the physiology necessary to allow a general understanding of the compartmental models and their control mechanisms is presented here.

Briefly, heat produced by biochemical reactions of cellular metabolism is transported by circulation of blood to the skin. There some of it is transferred to the environment normally by convection, by radiation, and by energy of vaporization of sweat. The parametric control in this case is by vasomotor control on the cutaneous arterioles and by sweat glands. A small amount of vaporization occurs in the lungs and by so-called insensible water loss (other than sweat) through the skin. As for water balance, which is closely regulated, feedback control is accomplished through the hypothalamus.

Though cellular metabolism is increased by secretion of thyroid hormone in order to add heat to a body placed in a cold environment, the major source of control heat is due to shivering and increased general activity of skeletal muscle. Feedback signals to the hypothalamus are obtained mainly from deep central and skin receptors [24, 25]. Other receptors are strategically located at various points such as in the respiratory tract.

5.5.2 Derivation of Model

Thermoregulatory models have been derived by conveniently lumping or discretizing the naturally distributed process in several different ways. For example, an energy balance may be performed on each lumped element. Anatomically, it is convenient and has been found to be surprisingly accurate for many studies to divide the body into three lumps—core, skeletal muscle, and skin. Again, it is assumed that the temperature and the properties of a given compartment are the same throughout that compartment. The core includes such internal organs as the lungs, heart, kidney, liver, and intestines.

As in Chapter IV, a heat balance on each compartment is used to derive a dynamical model. It is assumed that geometrically the system consists of three concentric cylinders with axial conduction negligible compared to radial condition. The latter assumption is a common one for systems of this sort. The model is formulated so that surface-to-volume ratios of each compartment conform with that of man himself. As for the body, the outer compartment corresponds to skin and the inner compartment to core.

Vasomotor activity is used to vary blood-flow patterns through core and skin, thus coupling skin to core. The blood has negligible mass heat capacity, and its heat transfer usually takes place in the form of convection. For simple modeling purposes, however, it may be assumed that this effect may be represented by variable tissue conductivity. Such representation is actually quite accurate since the blood circulation time within the body is very short (on the order of one minute) compared to the temperature response (on the order of several hours), and since the rate of heat transfer is nearly proportional to the temperature difference between core and skin. Consequently, vasomotor control is recognized as a core–skin conductivity term.

A heat balance on the core yields

$$c_{mc}\, dT_c/dt = q_b - q_{re} - q_{cm} - q_{cs}, \tag{5.49}$$

where T_c is core temperature, q_b is heat generated from basal metabolism, q_{re} is heat loss due to respiration and excretion, q_{cm} is heat transferred from core to muscle, q_{cs} is effective heat transferred from core to skin, and c_{mc} is core mass heat capacity. Heat transferred from the core to the muscle and from the core to the skin is described by

$$q_{cm} = (kA_{cm}/l_{cm})(T_c - T_m) \tag{5.50}$$

and

$$q_{cs} = (k_v A_{cs}/l_{cs})(T_c - T_s), \tag{5.51}$$

where k is effective conductivity; l_{cm} and l_{cs} are effective heat transfer lengths; A_{cm} and A_{cs} are effective heat transfer areas for appropriate compartments; $k_v(t)$ is equivalent vasomotor conductivity term which is a control variable; T_c is average core temperature; and T_m is average muscle temperature. While Eqs. (5.50) and (5.51) couple the core with other compartments, basal metabolism and respiration–excretion allow for some additive control, and vasomotor conductivity provides bilinear control.

A heat balance on muscles yields

$$c_{mm}\, dT_m/dt = q_s + q_{ex} + q_{cm} - q_s, \tag{5.52}$$

where T_m is average muscle temperature, c_{mm} is muscle core heat capacity, q_s is muscle shivering metabolism, q_{ex} is heat from muscle exercise, and q_{ms} is heat transferred to muscle from skin according to

$$q_{ms} = (kA_{ms}/l_{ms})(T_m - T_s). \tag{5.53}$$

Here, A_{ms} and l_{ms} are effective heat transfer areas and lengths, respectively.

Similarly, a heat balance taken at the skin shows that

$$c_{ms}\, dT_s/dt = q_{cs} + q_{ms} - (q_c + q_e + q_r), \tag{5.54}$$

where T_s is average skin temperature, c_{ms} is skin mass heat capacity, and q_c, q_e, and q_r are heat loss terms due to convection, evaporation, and radiation, respectively.

Again the system is described by a set of compartmental bilinear differential equations (5.49) to (5.56) with multiplicative and additive control terms. Compartmental temperatures T_c, T_m, and T_s are appropriate state variables for this thermoregulatory plant.

Numerous attempts have been made to model the feedback control equations which again make the closed-loop system more highly nonlinear or nonbilinear. Evaporation rate q_e may be approximated by [8]

$$q_e = \begin{cases} k_1(T_c - T_c^*), & \text{for } T_c > T_c^*, \\ q_{e0}, & T_c < T_c^*, \end{cases} \tag{5.55}$$

where T_c^* is desired core temperature (normal body temperature), q_{e0} is basal evaporation rate (almost zero), and k_1 is a positive constant. This model assumes that the corresponding response time is negligible compared to core temperature response. This equilibrium assumption is usually valid, but a more accurate function of system temperature is proposed by Crosbie *et al.* [26] and has the form of

$$q_e = \begin{cases} q_{e0} + k_4[k_2(T_b - T_b^*) + k_3(T_b - T_b^*)^4], & T > T_b^*, \\ q_{e0}, \quad T_b < T_b^*, \end{cases} \tag{5.56}$$

where k_2, k_3 are appropriate empirical constants, k_4 is an exercise constant, $T_b{}^*$ is a desired weighted body temperature, and

$$T_b = (l_s T_s + l_m T_m + l_c T_c)/(l_s + l_m + l_c),$$

where l_s, l_m, and l_c are effective thicknesses of the appropriate compartments.

Muscle shivering heat q_s and basal metabolic heat q_b are also nonlinear functions of core temperature and skin temperature, which may be approximated by linear equations; q_b is nearly constant, and q_s is approximated by [26]

$$q_s = \begin{cases} k_5(T_b{}^* - T_b), & T_b < T_b{}^*, \\ 0, & T_b > T_b{}^*, \end{cases} \tag{5.57}$$

where k_5 is a positive constant.

There seems to be some anticipation in the vasomotor control mechanism which suggests a proportional plus derivative feedback controller. Such controllers for a slab model have been derived from physiological data.

Of course, man may readily add heat, q_{ex}, by exercising and by varying his quantity and color of dress. The latter directly affects the q_c and q_r terms in (5.54). These terms are also nonlinear functions of skin temperature and environmental conditions.

5.5.3 State-Plane Analysis

It is convenient to examine a two-compartment model of thermoregulation since its behavior can be studied in a state plane. Skin and core are the compartments selected since temperatures in these compartments are most meaningful to system behavior and they are monitored by the central nervous system for regulation of temperature.

Similar to the derivation for three compartments, a heat balance on the core and on the skin yields equations of the form given by (5.49) to (5.51) and by (5.54). Or for two compartments in a more convenient form the model is

$$dT_c/dt = u_1(T_s - T_c) + v_1 \tag{5.58}$$

and

$$dT_s/dt = bu_1(T_c - T_s) + u_2(T_c - T_a) + v_2, \tag{5.59}$$

where

$$u_1 = \frac{k_v A_{cs}}{l_{cs} c_{mc}}, \qquad v_1 = \frac{q_b - q_{re} + q_{ex}}{c_{mc}}$$

$$u_2(T_c - T_a) + v_2 \approx \frac{q_c + q_e + q_r}{c_{ms}}, \qquad b = \frac{c_{mc}}{c_{ms}},$$

and T_a is ambient temperature; u_2, v_1, v_2 may be manipulated to some extent, but most of the discussion here will study the effect of $u_1(t)$, the vasomotor control term.

From Eqs. (5.58) and (5.59), the equilibrium state is defined by

$$\begin{aligned} T_{ce} &= (bv_1 + v_2)/u_2 + v_1/u_1 + T_a \\ T_{se} &= (bv_1 + v_2)/u_2 + T_a. \end{aligned} \tag{5.60}$$

Figure 5.2 shows the region S of possible equilibrium points with constraints of the form

$$\begin{aligned} 0 &< u_{1a} \le u_1 \le u_{1b}, \\ 0 &< u_{2a} \le u_2 \le u_{2b}, \\ 0 &< v_{1a} \le v_1 \le v_{1b}, \\ 0 &< v_{2a} \le v_2 \le v_{2b}. \end{aligned} \tag{5.61}$$

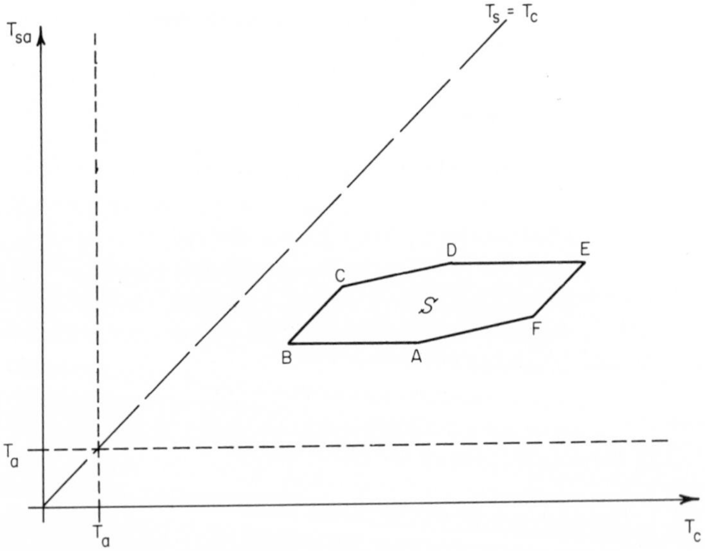

Fig. 5.2. Equilibrium region S for Eqs. (5.58) and (5.59).

Point A is obtained from (5.60) with $u_1 = u_{1a}$, $u_2 = u_{2b}$, $v_1 = v_{1a}$, and $v_2 = v_{2a}$, and the other points are similarly located as shown in Table 5.1. Line segments B–A and D–E locate two extreme equilibrium sets for $u_1(t)$, the main control of interest here.

TABLE 5.1

CONTROL VALUES FOR EQUILIBRIUM SET

Point[a]	u_1	u_2	v_1	v_2
A	u_{1a}	u_{2b}	v_{1a}	v_{2a}
B	u_{1b}	u_{2b}	v_{1a}	v_{2a}
C	u_{1b}	u_{2a}	v_{1a}	v_{2b}
D	u_{1b}	u_{2a}	v_{1b}	v_{2b}
E	u_{1a}	u_{2a}	v_{1b}	v_{2b}
F	u_{1a}	u_{2b}	v_{1b}	v_{2a}

[a] Point location in state space from (5.60); see Fig. 5.2.

Now assume that u_2, v_1, v_2 are held constant, in order to study controllability aspects of $u_1(t)$. In Fig. 5.3 a typical viability region is superimposed on an equilibrium set corresponding to extreme values of ambient temperature (T_{aa}, T_{ab}) and to extreme values of control (u_{1a}, u_{1b}). For example, $T_c > T_{cb}$ might cause irreversible damage to the nervous system; $T_c < T_{ca}$, core hypothermia causes "cold narcosis"; $T_s > T_{sb}$ results in skin burn; and $T_s < T_{sa}$ may cause freezing. Certainly, larger tolerances on skin temperature T_s can be admitted than those on core temperature T_c. The crosshatched region in Fig. 5.3 locates those admissible equilibrium states for survival with vasomotor control alone, which is the parametric or multiplicative control in this case. Note that altering the ambient temperature has a similar effect to adding clothing, exercising, or increasing metabolic rate according to this model. These take the form of additive control here. With vasomotor control alone, however, it is seen that the thermoregulator will maintain equilibrium viability for ambient temperatures on $[T_{aa}, T_{ab}]$.

It is of interest to study the state-plane behavior of the system (5.58) and (5.59) for extreme constant values of vasomotor control with the other control mechanisms constant. The eigenvalues of the system matrix in this case are described by

$$\lambda_{1,2} = -\tfrac{1}{2}[(b+1)u_1 + u_2] \pm \tfrac{1}{2}\{[(b+1)u_1 + u_2]^2 - 4u_1u_2\}^{1/2}.$$

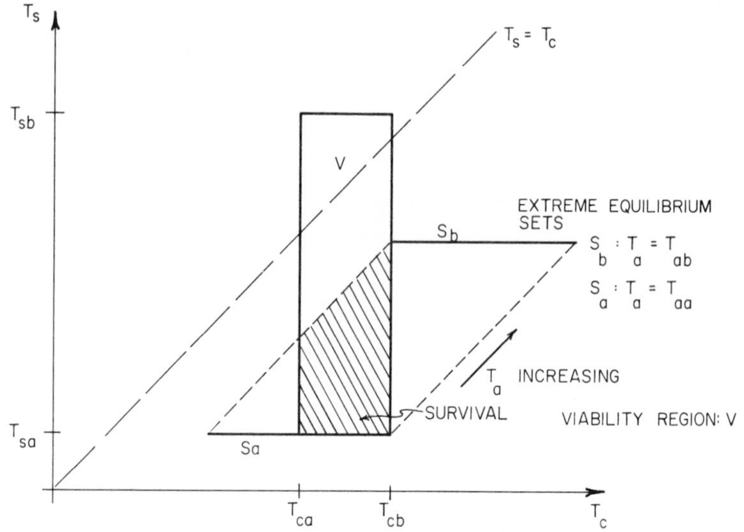

Fig. 5.3. Survival region with vasomotor control, $u(t)$.

From a Taylor series, the eigenvalues may be approximated by

$$\lambda_1 \approx -u_1u_2/[(b+1)u_1 + u_2]$$

and

$$\lambda_2 \approx -[(b+1)u_1 + u_2].$$

Therefore, equilibrium points for constant control are characterized as stable nodes with the characteristic behavior shown in Fig. 5.4 for two extreme equilibrium points, $\mathbf{x}_{ea}$ and $\mathbf{x}_{eb}$, corresponding to extreme controls u_{1a} and u_{1b}. The set of points S locates the equilibrium states for admissible constant controls.

From the solid trajectories for u_{1a} and the dashed trajectories for u_{1b}, the zone which can be reached from the admissible equilibrium S in finite time is defined by R, all points interior to the dotted trajectory segments. Likewise it is apparent that the incident zone, that is, the set of points from which S is reached in finite time, includes the entire finite state plane. Consequently, it is apparent that the region of complete controllability is defined by R also.

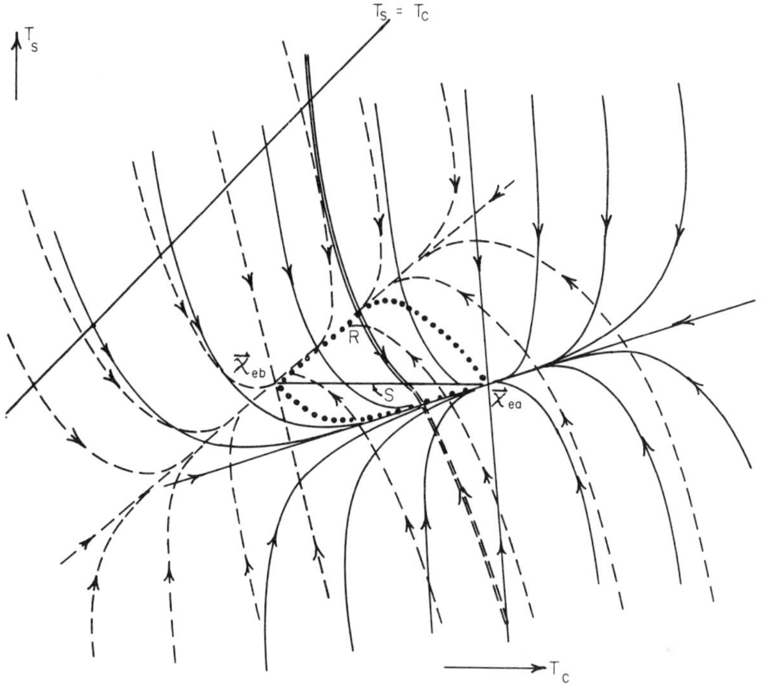

Fig. 5.4. Bilinear state portrait for bang–bang control u_{1a}, u_{1b}. Solid lines denote u_{1a}, dashed lines u_{1b}.

5.6 THE RESPIRATORY CHEMOSTAT

In this section, the respiratory system is regarded as a chemical O_2–CO_2 exchanger. Air breathed into the lungs is processed to provide oxygen to the body plant. Arterial blood, transporting oxygen to the body tissue, is in chemical equilibrium with the gas in the lungs. In the tissue, cellular metabolism converts O_2 to CO_2 which is transported back to the lungs by venous blood. The lungs, in turn, are continuously ventilated by breathing.

Physiologists have long recognized that this system is self-regulating. That is, if the body tissue for some reason becomes deficient in O_2 or overloaded with CO_2, the rate of ventilation of the lungs is automatically increased, thereby tending to increase the ratio of O_2 to CO_2 in the lungs and in the arterial blood. Associated with the O_2-to-CO_2 ratio in the blood is a certain acidity, or $[H^+]$ concentration. It has been shown that three agents, O_2, CO_2, and $[H^+]$, are each capable of exerting an independent effort on the ventilation rate, and hence are all controlled variables.

There has been an active effort over the past 25 years to describe this system in terms of mathematical control theory. In World War II, consideration of such practical questions as the oxygen requirements of pilots at high altitudes and the possible use of CO_2 to counteract anoxemia led Gray [27] to formulate a model for the "steady-state chemostat." This model assumed that the concentrations of O_2, CO_2, and $[H^+]$ in the arterial blood were the controlled variables, and hence involved three feedback loops. First, it was recognized that it is the chemistry of a respiratory center, located in the medullary portion of the brain, and not the chemistry of the arterial blood that actually regulates breathing. In steady state, however, these are simply related by constants.

For the first model of a dynamic chemostat derived in 1954, the average CO_2 concentration in body tissue was the controlled variable, and it was assumed that this was the same as the CO_2 concentration in the tissue of the respiratory center [10]. The agents O_2 and $[H^+]$ were ignored for simplicity, since they appear to be weaker stimuli than the CO_2 level. This model successfully predicts some features of transient response to CO_2 inhalation, but fails to account for the response to exercise, that is, to an increase in metabolic CO_2 production.

The third model, which was proposed by Grodins and James [28], is an elaboration of the above and includes an integral control mode in addition to the proportional control used previously in order to account for exercise phenomena. This model also allows the blood flow to the brain relative to total cardiac output to be a function of the CO_2 level in the respiratory center, corresponding to experimental observation.

5.6.1 Steady-State Chemostat

In order to describe the steady-state plant, two conservation equations can be written which must hold in steady state. First let pO_2, pCO_2 be arterial gas pressures and $[H^+]$ the arterial hydrogen ion concentration; $[pO_2]_s$, $[pCO_2]_s$, and $[H^+]_s$ are the reference settings for these quantities; pCO_2' and pO_2' are the partial gas pressures in the inspired air; MR is the metabolic rate of CO_2 production; and u is the lung ventilation rate. Now, the quantity $(pCO_2 - pCO_2')u$ is proportional to the net rate at which CO_2 flows out of the lungs, which must equal MR, the rate at which CO_2 is produced. With a similar argument for the net O_2 entering the lungs,

$$pCO_2 = pCO_2' + k(\mathrm{MR})/u, \tag{5.62}$$

and

$$pO_2 = pO_2' - k(\mathrm{MR})/u, \tag{5.63}$$

where k is a constant which includes barometric pressure. The quantity $[H^+]$ can be related to pCO_2 by a linear approximation

$$[H^+] = a(pCO_2) + b. \tag{5.64}$$

From empirical data on the responses to CO_2 inhalation, to disturbances in acid–base balance, and to anoxemia, Gray [27] was able to obtain a steady-state law for the controller of the form

$$\dot{V} = k_1[H^+] + k_2 pCO_2 + k_3(k_4 - pO_2) - k_5, \tag{5.65}$$

where $k_1, \ldots, k_5$, are positive constants.

It should be pointed out that the gains in the feedback loops, the numerical multipliers in (5.65), are certainly not universal constants. They can vary widely from person to person, and are not even constant for a given person under all conditions.

5.6.2 Dynamic Chemostat

This model was derived by Grodins [10] in order to account for the transient response to stepwise CO_2 inhalation. From test data, it is evident that the ventilation lags the arterial pCO_2 in time, and that the latter can show overshoot and undershoot. These facts suggest that it is the CO_2 concentration in body tissue that is fed back, since tissue has a large capacity for CO_2 storage.

This model ignores the agents O_2 and $[H^+]$, and assumes that the respiratory center can be lumped with all other tissues into a single homogeneous CO_2 reservoir with constant blood supply [27]. While the second assumption seems valid, there might be some question pertaining to the first. The accuracy of the model when compared to experimental data, however, does substantiate the model for at least the purpose served here. For this model, lungs and tissue form the two compartments, with x_2 and x_1 representing the CO_2 concentrations in tissue and lung, respectively. The analogy to an *RC* circuit is rather obvious. The tissue and lung CO_2 reservoirs correspond to capacitors, while the finite flow and pressure-dependent CO_2 absorption characteristics of the blood represent a resistance to the flow of CO_2.

Again, the system of equations may be derived from continuity equations on each compartment. Beginning with the lung reservoir, the CO_2 continuity equation states that the rate of change of lung CO_2 concentration $\dot{x}_1$ is equal to the difference between the rates of CO_2 inflow and outflow by all paths divided by the volume of the reservoir v_1. Thus,

$$dx_1/dt = (1/V_1)[uF^i(CO_2) + q_3 - q_2 - x_1u], \tag{5.66}$$

where $F^i(CO_2)$ is the fraction of CO_2 in the inspired air, q_3 is the rate at which CO_2 is delivered by venous blood, and q_2 and x_1u are the rates at which CO_2 leaves the lung via arterial blood and expired air, respectively. A similar CO_2 continuity equation can be written for the tissue reservoir, so that

$$dx_2/dt = (1/V_2)[\mathrm{MR} + q_2 - q_3], \tag{5.67}$$

where V_2 is tissue volume. It is assumed that the arterial blood is in chemical equilibrium with the air in the lungs. Then, the arterial CO_2 concentration can be expressed as a function of x_1 by means of a linearized CO_2 absorption curve, as

$$q_2/Q = B\alpha x_1 + \beta, \tag{5.68}$$

where Q is the cardiac output, B is barometric pressure, and α and β are the slope and intercept, respectively, of the linear absorption curve. We also assume that the venous blood is in equilibrium with the tissue CO_2 concentration, and that tissue and blood have the same CO_2 absorption curve. Then we have simply

$$x_2 = q_3/Q. \tag{5.69}$$

Finally, using (5.68) and (5.69) to eliminate q_2 and q_3 from (5.66) and (5.67), we obtain the equations of the controlled system:

$$dx_1/dt = (1/V_1)[u(F^i(CO_2) - x_1) + Q(x_2 - B\alpha x_1 - \beta)] \quad (5.70)$$

$$dx_2/dt = (1/V_2)[MR - Q(x_2 - B\alpha x_1 - \beta)]. \quad (5.71)$$

We observe that this system is bilinear. That is, the control variable u appears multiplying the state variable x_1. Of course, once the control law giving u as a function of state variables is substituted in the equation, the closed-loop system again becomes nonlinear.

The last assumption involves the nature of the control law. First, it is assumed that

$$u = k_p x_2 + c, \quad (5.72)$$

that is, a proportional control of x_2, where k_p is the controller gain. We have now a complete mathematical description of the system, in terms of certain physiological parameters such as lung and tissue volumes V_1 and V_2, cardiac output Q, blood CO_2 absorption parameters α and β, and controller gain k_p, all of which are assumed to be measurable constants. Equations (5.70) and (5.71) determine the response of the state variables x_1 and x_2 to arbitrary disturbances $F^i(CO_2)$ and MR. Grodins [10] gives an analysis of the stability of this system but this is unnecessary if we observe the complete analogy with an *RC* circuit.

The interesting features of system solutions and experimentation are that the response time in the $u(t)$ response is a function of $F^i(CO_2)$, the driving amplitude, and that the percentage overshoot in the arterial pCO_2 is also a function of $F^i(CO_2)$, being much larger for small $F^i(CO_2)$. However, if the system is driven by MR, that is, by exercise, the agreement is not so good. For one thing, this system will always show steady-state error since proportional control is used. However, the arterial pCO_2 error has been shown to be zero in exercise. For such reasons another feedback control model was proposed [28, 29].

In the previous models the respiratory cycle was ignored and u was considered to be a continuous flow. However, u is in fact cyclic and the arterial pCO_2 oscillates about a mean with frequency equal to the breathing frequency. It has been shown that although mean arterial pCO_2 remains constant during exercise, the amplitude of oscillations about this mean increases. Also, it is suggested that signal information residing in this altered dynamic behavior might contribute to the regulation of breathing in exercise [30].

In order to include the respiratory cycle in the controlled system, the equations may be modified as follows. If a sinusoidal respiratory cycle is assumed with

$$u = \tfrac{1}{2}\omega V_0 \sin \omega t, \tag{5.73}$$

where V_0 is tidal volume and ω is the frequency of breathing, then instead of (5.66) for the CO_2 continuity in the lungs, there are now two equations:

$$d(Vx_1)/dt = uF^i(CO_2) + q_3 - q_2 \tag{5.74}$$

for the inspiratory half-cycle, and

$$d(Vx_1)/dt = ux_1 + q_3 - q_2 \tag{5.75}$$

for the expiratory half-cycle.

A second modification is made to account for the experimentally observed fact that the fraction of total cardiac output flowing to the brain is actually a strong function of the CO_2 concentration in the brain. Defares [31] has shown, by means of an analog computer, that this modification improves the agreement between the model and biological responses for stepwise CO_2 inhalation. For, without this modification, the percentage undershoot in the "off" transient is large, at least 75 percent for any $F^i(CO_2)$, while for some human subjects the experimentally observed undershoot can be much smaller than this. With the modification, the percentage undershoot depends on the cardiac output of the subject, and agrees very well with experiment. Thus it is assumed that the blood flow to the brain is $QF(x)$, where F is the fraction flowing to the brain and is a function of x, the CO_2 concentration in the brain. Finally, since in exercise the heart rate increases along with other muscular activity, the cardiac output is assumed to be a linear function of total metabolic rate according to

$$\tau\dot{Q} + Q = k\text{MR}. \tag{5.76}$$

To complete the mathematical description of this system, the CO_2 continuity equation for lumped tissue, Eq. (5.67), needs to be replaced by two such relations, one for the brain exchanger and one for the non-brain exchanger. Also, an empirical law which gives $F(x)$ is needed.

Grodins and James [28] suggest that the controller receives an input signal from the brain, pCO_{2B}, and from this generates a continuously operating proportional control. It operates on the mean absolute value of the time derivative of the arterial pCO_2 and on the existing average magnitude of ventilation rate. Each of these signals is compared with a reference and multiplied by a gain constant, and the products are then added. If the sum Y is negative, it is simply added to the proportional signal from the brain CO_2 to generate the final output. If, however, Y is positive, its integral is added to the proportional brain tissue signal. It turns out that this decision process is dictated by the relationship between u and average arterial pCO_2 time derivative magnitude. In exercise these two quantities maintain the same relative size as they have at rest. The controller thereby senses that the CO_2 load is a metabolic one and the integrator operates to zero the steady-state error in mean arterial pCO_2. In CO_2 inhalation, however, the average arterial pCO_2 time derivative magnitude is relatively small. The controller thereby senses that the CO_2 load is an external one, and the integrator is bypassed. Grodins and James [28] give specific empirical equations which describe the operation of this controller.

Grodins and James [28] use an analog computer to study the responses of this model under various input conditions of exercise and CO_2 inhalation, and obtained very "lifelike" behavior. As they put it, the agreement is almost "embarrassingly close." Of course, this controller was invented to correspond to the experimentally observed phenomenon of respiratory response to either CO_2 inhalation or exercise.

5.7 THE CARDIOVASCULAR REGULATOR

Control of the cardiovascular system is somewhat similar to, but considerably more complex than, the respiratory chemostat. Still, the system can be compartmented and conservation questions can be derived to describe the compartmental dynamics. Again, the ensuing model is approximately bilinear, with compartmental blood pressures as state variables and resistances to blood flow in peripheral arterioles as multiplicative controls. For example, these resistances are altered by the vasomotor nerves during exercise, hemorrhage, or shock.

The cardiovascular plant being controlled is comprised of a mechanical section (heart and vascular networks) and a chemical exchanger (tissue). The controller consists of neural and endocrine sections which are activated by error signals received through cardiac and vascular centers in the medulla in response to pressure feedbacks through arterial baroceptors.

The primary controlled variables of the system are the levels of tissue (venous) pO_2 in the several parallel systemic circuits. It is assumed that other tissue agents, such as pCO_2, pH, and metabolites, are represented indirectly through pO_2.

Within the plant, the heart provides cardiac output $w(t)$ as an input to the networks. Arterial and venous pressures are fed back to the heart, and local systemic flows feed the tissue gas exchanger. Tissue (venous) pressures, pO_2, feed back to effect systemic circuit pressures and local flows through manipulation of systemic arteriolar resistances. Network pressures are fed back to vary cardiac output. The neural signals to the heart manipulate cardiac output through heart frequency, contractile strength, and filling impedance. Neural signals, also, manipulate arteriolar resistance and venous compliance, and activate the adrenal medulla for release of epinephrine and norepinephrine. These in turn effect the heart and circuits.

Following Grodins [10] the mechanical portion of the cardiovascular system may be divided into four compartments, with $v_2(t)$, cardiac output from the right heart, and $v_1(t)$, cardiac output from the left heart, as inputs to an arterial and a venous compartment, respectively. The four conservation equations are

$$c_1\, dx_1/dt = u_1(x_2 - x_1) + v_1, \tag{5.77}$$

$$c_2\, dx_2/dt = u_1(x_1 - x_2) - v_2, \tag{5.78}$$

$$c_3\, dx_3/dt = u_2(x_2 - x_3) + v_2, \tag{5.79}$$

and

$$c_4\, dx_4/dt = u_2(x_3 - x_4) - v_1, \tag{5.80}$$

where $x_1, \ldots, x_4$ are compartmental pressures in the arteries filled from the left heart, veins to the right heart, arteries filled from the right heart, and veins to the left heart, respectively; $c_1, \ldots, c_4$ are corresponding compartmental capacities (assumed constant); and $u_1(t), u_2(t)$ are inverse pulmonary–arterial and pulmonary–venous network resistances to flow. Since it is a closed system, pressures are constrained so that

$$c_1x_1 + c_2x_2 + c_3x_3 + c_4x_4 = B,$$

where B is a positive constant.

Cardiac outputs obtained from the left and right hearts are additive control functions with complicated nonlinear feedback approximated by [10]

$$v_1 = k_4x_4/(k_1x_1 + k_5)$$

and

$$v_2 = k_2x_2/(k_3x_3 + k_5).$$

It must be emphasized that only a very crude partial model has been studied here. More details of this model are presented by Grodins [10]. Certainly more work needs to be done to include such complications as chemical feedback effects of pO_2 on systemic resistance, effect of vasomotor signals on capacity of veins, fluid exchange across capillaries, and the presence of venular resistance, to mention a few.

In summary, cardiovascular regulation does involve a variable structure which may be approximated by a bilinear system as a result of multiplication of pressures by controlled resistance terms or susceptances. Cardiac output is an additive control in the model. Cardiac output increases to maintain arterial pressure x_1 with decreasing disturbances in systemic resistance due, for example, to hypoxia, anemia, exercise, or heat. Also, resistance elements are adjusted to maintain arterial pressure x_1 with low cardiac output from such disturbances as gravity stress, shock, and hemorrhage.

5.8 OTHER BILINEAR SYSTEMS

Since the law of mass action under influence of a catalyst, such as an enzyme for the cellular plant, can be modeled effectively as a bilinear system, it is expected that other biological processes which are described by similar reactions can be represented by a bilinear system.

5.8.1 Thyroxin Regulation

The regulation of circulating thyroxin in the body is one such process which has been modeled recently by a bilinear system [32]. Thyroxin is a major thyroid hormone which is bound in plasma to certain protein molecules. The reaction between these hormones and their binding protein is reversible, and consequently there is free hormone, free protein, and protein-bound hormone molecules present in the blood, and in a state of interchange. For this process, the law of mass action yields a bilinear model similar to (5.4) and (5.5), where x_1 is free thyroxin concentration, x_2 is protein-bound thyroxin concentration, and u is free protein concentration. Again, complex feedback controllers through the hypothalamus have been hypothesized [32].

5.8.2 Beef Model

Also similar to the cellular model is the bilinear model for a beef animal, which has been used successfully by Bacon and Witz [33]. If x_1 is the beef weight and x_2 is $\dot{x}_1$, the time rate of change of beef weight, then the process may be described by

$$dx_1/dt = x_2$$

and

$$dx_2/dt = ax_2 + u(ax_1 + x_2) - cau, \tag{5.81}$$

where u is total energy consumed minus energy required for maintenance, a is a constant, and c is a hereditary constant.

5.9 SUMMARY

Compartmental bilinear physiological systems are analyzed in this chapter along with system realization by means of tracer experiments. The techniques developed are applied to body water balance. Other physiological models of this category include the respiratory chemostat, the cardiovascular regulator, thermoregulation, and biochemical cellular processes. The compartmental models are derived from conversation equations of one type or another. In all cases parametric or multiplicative control evolve in a natural manner.

Bilinear models arise from the law of mass action for chemical reactions with a catalyst as a control. For cellular biochemical reactions, enzyme concentration is the catalyst. This represents a very general class of processes.

The work presented here only begins to scratch the surface of biological systems analysis. Certainly, more new problems have arisen than have been solved. But, it does show what can be done under certain conditions and assumptions.

Mathematical modeling in physiology may some day play an important role in diagnosis. Models may be stored and used for comparison when the patient becomes ill. Simulation may help pinpoint the diagnosis and even the treatment. The effects of certain drugs on system parameters may be obtained by simulation. Inverse optimization, deterioration of performance, and loss of system controllability may some day be an important part of diagnosis.

A controllability analysis made in this chapter does provide analytical insight into thermal regulation. Again, as first explained in Chapter II, the purpose of multiplicative or parametric control for extending the region of controllability about given equilibrium points is obvious. Likewise, additive control is necessary to reach a broad range of equilibrium points. These aspects are useful for applications in physiology as well as engineering. Still, it is unfortunate that such precise theoretical results as the sufficiency conditions for complete controllability have not been useful for these problems.

Again, the significance of the bilinear model with independent additive and multiplicative control is apparent. Applications of certain theoretical results to these special problems are left for the reader. Some problems for this purpose are given in the exercises.

The role of controllability and its results for linear systems are shown to be quite useful for the realization of bilinear models to describe the behavior of compartmented physiological processes. This work is made useful experimentally by means of radioactive tracers. Compartmental access and geometry are related to controllability as a general basis for modeling and identification of compartmented bilinear systems by means of tracers.

Exercises

5.1 (a) Show that the population model with feedback control (5.1) does constrain population.
(b) Does this feedback control suggest that a quadratic performance index may be optimized?

5.2 Can the biochemical process (5.4) and (5.5) be completely controllable with finite enzyme concentration? Why?

5.3 (a) Discuss the use of compartmented models and tracer experiments for modeling and identification of systems in general.
(b) Compare compartmentation and discretization for anatomical and mathematical convenience.

5.4 Derive a linearized model for active transport from (5.11).

5.5 Put (5.13) into the form of (5.15). Identify any useful properties of the system matrices.

5.6 Distinguish between "steady state" and equilibrium as used in physiology and in system theory.

5.7 Why is the tracer system (5.17) sometimes said to be adjoint to the original system (5.15)?

5.8 Why are minimal realizations and controllability so important to system modeling?

5.9 (a) Derive a linearized transfer function for changes in body core temperature taken about normal equilibrium temperature with respect to appropriate changes in the effective vasomotor conductivity term.
(b) Derive the same with respect to changes in heat generated from basal metabolism.
(c) Compare the effectiveness of these two control variables for body temperature control.

REFERENCES

1. Pielou, E. C., "An Introduction to Mathematical Ecology." Wiley (Interscience), New York, 1969.
2. Markert, C. L., Biological limits on population growth. *BioScience* **16,** 859–862 (1966).

3. Green, D. E., and Hatefi, Y., The mitochondrion and biochemical machines. *Science* **133**, 13–19 (1961).
4. Eigen, M., and Hammes, G. G., Elementary steps in enzyme reactions. *Advan. Enzymol.* **25**, 1–38 (1963).
5. Fried, J., Compartmental analysis of kinetic processes in multicellular systems: A necessary condition. *Phys. Med. Biol.* **13**, 31–43 (1968).
6. Hill, T. L., and Kedem, O., Studies in irreversible thermodynamics III. Models for steady state and active transport across membranes. *J. Theoret. Biol.* **10**, 399–441 (1966).
7. Alvi, Z. M., Predictive aspects of monitored medical data. Ph.D. dissertation, School of Engrg., Univ. of California, Los Angeles, 1968.
8. Milsum, J. H., "Biological Control Systems Analysis." McGraw-Hill, New York, 1966.
9. Snyder, W. S., Fish, B. R., Bernard, S. R., Ford, M. R., and Muir, J. R., Urinary excretion of tritium following exposure of man to HTO—a two exponential model. *Phys. Med. Biol.* **13**, 547–559 (1968).
10. Grodins, F. S., "Control Theory and Biological Systems." Columbia Univ. Press, New York, 1963.
11. Langer, G. A., and Brady, A. J., The effects of temperature upon contraction and ionic exchange in rabbit ventricular myocardium. *J. Gen. Physiol.* **52**, 682–713 (1958).
12. Cannon, W. B., "The Wisdom of the Body." Norton, New York, 1939.
13. Sheppard, C. W., and Householder, A. S., The mathematical basis of the interpretation of tracer experiments in closed steady-state systems. *J. Appl. Phys.* **22**, 510–520 (1951).
14. Sheppard, C. W., The theory of the study of transfers with a multicompartment system using isotopic tracers. *J. Appl. Phys.* **19**, 170–176 (1948).
15. Berman, M., and Schoenfield, R., Invariants in experimental data on linear kinetics and the formulation of models. *J. Appl. Phys.* **27**, 1361–1370 (1956).
16. Chen, C. T., "Introduction to Linear System Theory." Holt, New York, 1970.
17. Lancaster, P., "Theory of Matrices." Academic Press, New York, 1969.
18. Hearon, J. Z., Theorems on linear systems. *Ann. N.Y., Acad. Sci.* **108**, 36–68 (1963).
19. Smith, W. D., Compartmental bilinear models and tracer techniques in the analysis of biological control systems. Ph.D. dissertation, College of Engrg., Univ. of Oklahoma, Norman, 1972.
20. D'Angelo, H., "Linear Time-Varying Systems: Analysis and Synthesis." Allyn & Bacon, Rockleigh, New Jersey, 1970.
21. Ruch, T. C., and Patton, H. D., "Physiology and Biophysics," 19th ed. Saunders, Philadelphia, Pennsylvania, 1965.
22. Pitts, R. F., "Physiology of the Kidney and Body Fluids." Yearbook Publ., Chicago, Illinois, 1963.
23. Smith, H. W., "Principles of Renal Physiology." Oxford Univ. Press, London and New York, 1956.

24. Hardy, J. D., Physiology of temperature regulation. *Physiol. Rev.* **41**, 541–606 (1961).
25. Granit, R., "Receptors and Sensory Perception." Yale Univ. Press, New Haven, Connecticut, 1955.
26. Crosbie, E. J., Hardy, J. D., and Fessenden, E., Electrical analog simulation of temperature regulation in man. *IRE Trans. Bio-Med. Electron.* **BME-8**, 4 (1961).
27. Gray, J. S., The multiple factor theory of respiratory regulation. *Science* **103**, 739 (1946).
28. Grodins, F. S., and James, G., Mathematical models in respiratory regulation. *Ann. N.Y. Acad. Sci.* **109**, 852–868 (1963).
29. Defares, J., The stability of the respiratory servomechanism: An analog computer study. *Progr. Cybernet.* **1** (1964).
30. Yamamoto, W. S., and Brobeck, J. R., "Physiological Controls and Regulations." Saunders, Philadelphia, Pennsylvania, 1965.
31. Defares, J., A cybernetic analysis of the respiratory chemostat. *Symp. Physico-Math. Aspects of Bio., Internat. School Phys., Varenna, Italy, July 1960.* Academic Press, New York, 1962.
32. Distefano, J. J., III. A model of the regulation of circulating thyroxin, unbound and bound to plasma proteins and its response. *Math. Biosci.* **4**, 137–152 (1969).
33. Bacon, C., and Witz, J. A., Oklahoma State Univ., Norman, private communication, February 1971.

VI

Socioeconomic Systems

6.1 CONCEPTS OF SOCIOECONOMIC MODELS

There is an obvious need for systems analysis in socioeconomics, and the demands of modern society will lead to a large research effort in this sector. One needs only to scan the modern systems literature to realize that such research is already emerging. There have been numerous special journal issues, conferences, and comprehensive reports devoted to various facets of urban dynamics, public systems, health care, traffic control, and economic modeling.† Of these topics, economic modeling probably has received the earliest and most concentrated effort. Keynes [1] is credited with the earliest mathematical studies in economics in which he studied periodic deviations from a presumed normal equilibrium state. This work led to the development of the so-called

† For example, see (1) Special Issue on Urban and Public Systems, *IEEE Trans. Syst. Sci. Cybernetics*, SSC-6, No. 4, 1970; (2) Future Fields of Control Application, *Proc. NASA Symp.*, MIT, Feb. 10–11, 1969; (3) Interdisciplinary Res. Topics in Urban Engrg., Rep. of Urban Engrg. Study Committee, ASEE, Washington, D.C., 1969.

Keynesian theory which includes roughly the following necessary conditions for both short-run static equilibrium† and for long-term growth equilibrium in an industrial economy.

1. Short-term static equilibrium requires that total savings equal planned investment.

2. Long-term growth equilibrium requires that the investment in a new plant and equipment of one period be matched by a corresponding increase in production in the following period [2].

Similar conclusions were reached independently by Harrod [3] and Domar [4]. Another fundamental piece of research on economic modeling was conducted by Frisch [5] who showed that economic cycles can evolve by the economy's inherent structure independent of external inputs. Another milestone in economic systems was set by Samuelson in 1939 [6]. As time progressed numerous other significant papers appeared under the heading of economic models, macroeconomic analysis, or econometrics. They include the work of Metzler [7], Phillips [8], Goodwin [9], and Duesenberry [10].

While at first glance this work is quite dispersed it does have a common thread which ties in nicely with the bilinear models presented here. Multiplicative control, such as that which appears from capital investment, is a dominant factor. Generally, it has been assumed that the models are linear in state. Rather than analyze the bilinear model, however, researchers have gone to linear models since linear system theory has been so completely developed. As noted by Runyan [2], it is very likely that gross simplifications to a complicated problem have hindered rather than helped interdisciplinary work on economic systems. This is natural since linear models cannot satisfy well-developed intuitive concepts in economics except for special problems. The bilinear model, with its parametric control or economic investment parameter, offers a significant improvement in this respect.

Still, such a conclusion does not imply that all the results based on linear economic models should be discarded. Linear, time-invariant, systems analyses have been quite effective on a piecemeal basis. Again, this is consistent with the results developed in Chapters II and III which

† While such terms as equilibrium and steady state have precise meanings in system theory, they are used in slightly different contexts in applications. In both Chapters V and VI, the author finds it more meaningful to use the latter approach.

show the role of bang–bang or piecewise constant control and the role of controllability. Repeating these results for convenience, bang–bang control is frequently a candidate for the optimal strategy, and under such control the bilinear process behaves as a piecewise linear system with constant coefficients. The controllability analyses presented in Chapter II further substantiate the role of bilinear economic models since the apparent lack of controllability of linear systems and the high degree of controllability of bilinear systems by means of controlled implementation of unstable modes (or equilibrium growth modes in an economic sense) are established there.

6.2 URBAN DYNAMICS

Urban modeling, another aspect of socioeconomic systems, has been studied recently by numerous researchers. The most famous of this work has been reported by Forrester [11, 12], who has gone a step further in an attempt to use his work as a basis for mathematical modeling in general [13]. Like the early work on economic modeling, this has been an impetus for continued study, and it provides some very general explanations of urban growth and development. Again, however, the modeling relies mainly on linear system theory and does not utilize the more modern concepts of linear systems, such as the concept of state. Just as parametric control appears in the form of bilinear control for biological growth and for economic growth it may be used to provide more accurate descriptions of urban dynamics.

An example of parametric control in modern dynamics is the bilinear model presented by Jaeckel [14] to modify Forrester's model [12] for industrial regeneration. For this process the new enterprise construction is described by the following algebraic equation which fits into the overall dynamical model

$$y_c = y_{cd}u_{cr} + v_{br} + v_{ir}, \tag{6.1}$$

where y_{cd} is new enterprise construction demand, u_{cr} is labor construction restriction, v_{br} is mature business revival, and v_{ir} is declining industrial renewal. While the latter two variables appear as additive controls, they may be further broken down into a mature business state term multiplied by a business revival multiplier and into similar terms for declining industry. Hence labor construction restriction, business revival multiplier, and declining industry revival multipliers are all multiplicative or bilinear controls.

For a population transfer model, Gibson [15] utilizes

$$dy_{\mathrm{ua}}/dt = (y_{\mathrm{u}} + y_{\mathrm{e}})u_{\mathrm{mp}}, \tag{6.2}$$

where y_{ua} is the number of underemployed arrivals, y_{e} is the number of laborers, y_{u} is the number of underemployed laborers, and u_{mp} is the attractiveness for migration multiplier. In the overall system the latter control is derived by a multiplication of more fundamental quantities—underemployed arrivals' mobility multiplier, underemployed housing multiplier, public expenditure multiplier, underemployed housing program multiplier. These multipliers are in turn derived from more basic parameters, which are underemployed mobility, underemployed per housing ratio, tax per capita ratio, underemployed per job ratio, and underemployed housing program rate, respectively. The former multipliers are all multiplied together to generate so-called attractiveness for migration multipliers which becomes an input to a first-order linear system with a 20-year perception time constant. The output is the attractiveness for migration multiplier perceiver. Again, any controlled manipulation of perception time constant would result in a bilinear subsystem.

As discussed in previous chapters, bilinear models and their related theory are most effective for the analysis and simulation of basic system components such as those shown here for the generation of underemployed population arrivals in a population subsystem of an urban model. Certainly the overall model structure is nonbilinear and quite complex with components connected together and parametric controls generated from state variables, such as population of underemployed arrivals in a given year.

Major complex subsystems in Forrester's model are business activity, housing, and working population [12]. These are recognized as more basic aspects of urban growth and stagnation than are such processes as city government, social culture, or fiscal policy. While this assumption is questionable and effective urban models might be perceived from other basic subsystems, Forrester does simulate progressions for these three subsystems. With models such as those discussed above for underemployed population, maturation, recession, and destruction of industry are simulated. Correspondingly, luxury houses are built and deteriorate to become workers' houses which further deteriorate into slum housing and eventually are demolished. Labor population arrives and departs from the urban area, is downgraded to become underemployed or unemployed, or is promoted.

Any analysis which involves such broad socioeconomic aspects as Forrester's work is bound to be wide open to criticism. Some excellent criticisms are presented in the special issue on "Urban Dynamics: extension and reflections" in the *IEEE Trans. Syst., Man and Cybernetics,* **SMC-2** (1972). While the system scientists are critical of his mathematical assumptions (on structure, etc.), the socioeconomists are critical of his assumptions in their sector. Still, it must be granted that this pioneering effort is establishing a base for more effective studies.

Similar socioeconomic system studies are now being conducted on Oregon's Willamette Valley (by Oregon State University), on a region of Wisconsin (by the University of Wisconsin), on Eastern Tennessee (by Oak Ridge National Laboratory), on Nigeria (by Michigan State University), and on a region of Vancouver (by the University of British Columbia)—to mention a few.

For the agricultural sector, which is modeled in the Nigerian study [16], the production of crops is manipulated parametrically by the controlled use of fertilizer. In ecology, a similar effect is realized by clear cutting in the forests or by the use of defoliants in the jungle.

Just as in Forrester's work, numerous simplifying assumptions must be made in order to generate a manageable model. The model and its simulation should provide insight into future development, and it should allow better overall systems planning in industrial development, housing, transportation, and health care. Still the distinction between model and actual society must be realized.

While controllability and optimal control should eventually be useful in the analysis of socioeconomic systems, the more basic work is needed to develop sufficiently accurate socioeconomic models. Consequently, the application of the more theoretical results provided in Chapters I to III would be more academic than useful at this point.

The remaining portion of this chapter is devoted to a fundamental socioeconomic case study of a fascinating nation, the Peoples Republic of China. Unfortunately, the data available are quite sketchy so that the analysis is very superficial and philosophical. It is hoped that this cursory study will provide a base for more intensified research. The discussion follows that of Mohler [17], and the application of bilinear models is discussed throughout. In the demographic analysis, an interesting relation is shown to bilinear compartmental models in physiology. The theory established in Section 5.3 has obvious generalizations.

6.3 BACKGROUND FOR CASE STUDY

To better understand Sino foreign policy and the strong nationalism of China's people which effect her economic policy, it is worthwhile to glance back into Chinese history. This history has involved cycles of power and glory interspersed with periods of civil strife and disunity. Throughout all history, Chinese civilizations have experienced cycles of rise and fall as presented in Table 6.1. Again the behavior is typical of that of a bilinear system with alternating, positive and negative, policy. For example, national expansion, peace, and prosperity were dominant during much of the reign of the Han, Tang, and Ming Dynasties. Between periods of resurgence, however, complete disunity and revolution seemed to be the rule. As history repeats, it appears that the world is witnessing renewed Sino resurgence again. For after World War II, civil war erupted in China, and the Communists were victorious. The country was finally united after 250 years of chaos. It is suggested in the analysis presented here that China has been enduring a state of transition to national expansion and unification—consistent with history, but not without small oscillations of the kind shown throughout history on a larger scale and so conveniently modeled by a bilinear system. The tight control of its masses exerted through the government structure of the People's Republic of China (PRC) may just be the mechanism to accomplish this resurgence.

6.3.1 Population

In area, China is the third largest country in the world, but in population she is by far the world's largest. China's 1973 population, based on the 1953 census, the simple bilinear model given by Eq. (1.3), and the

TABLE 6.1

HISTORICAL SINO POWER OSCILLATIONS[a]

Approximate Time period	Dynasty	Events	
		Resurgence	Recession
200 B.C.–A.D. 200	Han	Expansion, prosperity, cultural development, peace	—
A.D. 200–600	San Kuo, West Chin, East Chin, Six dynasties, Suei	—	Rise of northern tribes, removal of capital from south, disunity, war
600–1100	Suei, Tang, Wutai, North Sung	Engineering feats, expansion, prosperity, peace, cultural development, rise of north	—
1100–1400	North Sung, South Sung, Yuan	—	Removal of capital from North, disunity, civil war
1400–1850	Ming, Manchu	Engineering feats, unity, peace, prosperity, cultural development	—
1850–1950	Manchu (Ching), Republic	—	Loss of capital, foreign encroachment, famine, war
1950–	Communist government	Engineering feats, signs of unity and prosperity	—

[a] See Lee [18].

coefficient given in Fig. 6.1, is approximately 820 million. Every ten years, the Chinese population increase is nearly equal to the total population of the United States—over 200 million in 1973. The model predicts that China will have over a billion people in this decade. The potential power of China really lies in her masses. Before 1950, China's economy sagged while her population rapidly increased. Indeed, the large population was a hindrance to fragmented China—a China that could not properly maintain the schools, hospitals, and factories she already possessed, yet alone build new ones to cope with her ever growing population.

While the United States is well proportioned with about six percent of both the world's land and the world's people, China is extremely unbalanced with nearly a quarter of the world's population but less than seven percent of the world's land area. Also, it is interesting to note that the Soviet Union with only nine percent of the world's population (about 280 million in 1972) has over 15 percent of the land. The Soviets have an average population density of only about 30 people to a square mile while China has about 210 people per square mile.

Even more pronounced than in the United States is China's concentration of people along her east coast. There, the present population densities vary from over 1100 people per square mile in Shantung and Kiangsu (which does not include Shanghai's municipality of over ten million) to about 320 people per square mile in Fukein (across the straits from Taiwan). About 26 percent of the Chinese live in the coastal provinces, within 200 or 300 miles of the sea. Over 15 percent of the people probably live within 60 miles of the sea. As far west as the plains northwest of Chungking, the countryside population density is at least 1000 people per square mile. Between east and west, the fertile Yangtze and Yellow River basins, along with their tributaries, are extremely densely populated. While this information may seem useless to the systems theoretician, it is of significance to the socioeconomic planner or modeler.

Only about 3.5 percent of the population presently lives in the four western provinces (Sinkiang, Kansu, Tsinghai, and Tibet) that compose 42 percent of the land area.† Further, if the outer provinces to the north (Inner Mongolia and Heilungkiang) are excluded, it is recognized that

† Data presented here have been computed by the author from population density maps and from population data that are readily available in the Encyclopedia Brittanica as well as in other encyclopedias.

93 percent of the population is contained in about 1.52 million square miles, or the Chinese average population density over an area about half the size of the United States (without Alaska and Hawaii) is nearly 500 people per square mile.

In 1953, only 13 percent of the Chinese masses lived in the largest 1600 urban areas. This is a striking contrast to the Soviet Union where 13 percent of the people live in only eight urban areas. The urban population characteristics for the United States are similar to those of the Soviet Union. As China's industry grows, more people will move to metropolitan areas. In fact, by 1957 it was estimated that 14 percent of the population was urban. It is doubtful though that such a movement will drastically affect the overall population distribution in the near future. Many eastern municipalities are already overpopulated. Shanghai, for example, is extremely crowded with about 23,000 people per square mile of city, and with about 7000 people per square mile of urban area (772 square miles). Probably some coastal areas (such as Peking, Shanghai, and the Mukden-Dairen area), however, will continue to expand radially in a manner similar to the Boston to Washington area which holds about 2000 people to a square mile, but China has declared her intentions to limit urban population to 110 million.

As China progresses, it appears that a great western and northern migration will take place. Eventually such a western migration may dwarf the American move westward. A clue that such a migration is already evolving may be obtained from the estimated population increases for the various provinces from 1953 to 1957. For example, while the total population increased ten percent, Inner Mongolia's population surged by 50 percent, Sinkiang's population increased 36 percent, Heilungkian's population (on the Soviet border) by 25 percent, and Tsinghai's population by 22.4 percent in just four years. Only the great industrial provinces of Liaoning (in Manchuria) with 30 percent increase and Hopeh (surrounding the municipality of Peking) with 18 percent increase showed comparable population growth.

It is interesting to note the analogy between bilinear compartmental models in physiology (see Section 5.3) and in demography. Here compartments may represent geographical regions or provinces. Migration rates, birth rates, and death rates are analogous to intercompartmental fluxes, compartmental substance generation, and compartmental substance destruction. Also tracer techniques can be used for demographic modeling just as for physiological modeling.

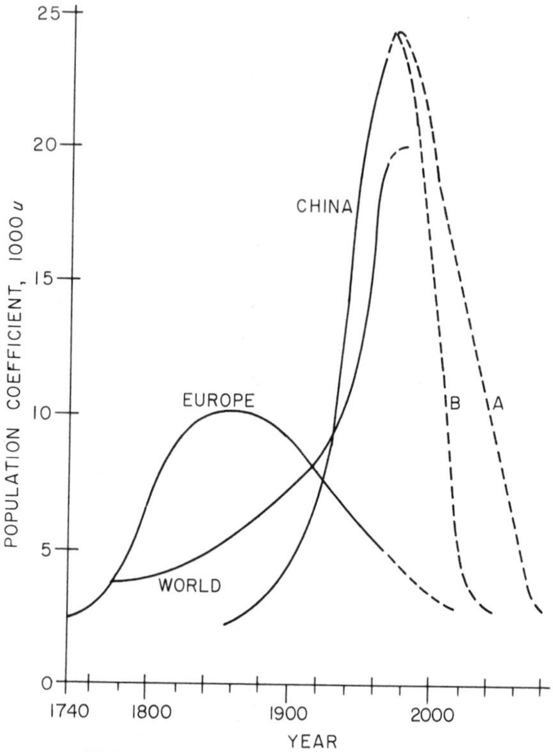

Fig. 6.1. Population coefficient, $u(t)$.

Though the 1953 Sino census (583 million) is generally accepted, there are conflicting views that on one hand, the estimate is too small (see Ho [19]), and on the other, slightly large (see Taeuber [20]). In any event, the estimate is reasonably consistent with figures published in the eighteenth and nineteenth centuries. For example, population figures for 1775 and 1850 were 265 million and 430 million.

The population coefficient u, given with (1.3) is an important index of performance for a nation's standard of living. As China advances economically, disease and famine continue to decrease, and the population coefficient continues to rise. Generally, however, birth control plays a constraining role as the economy rises and people become more educated. Emerging societies have experienced this process throughout history, and Fig. 6.1 presents the population coefficient for Europe, the

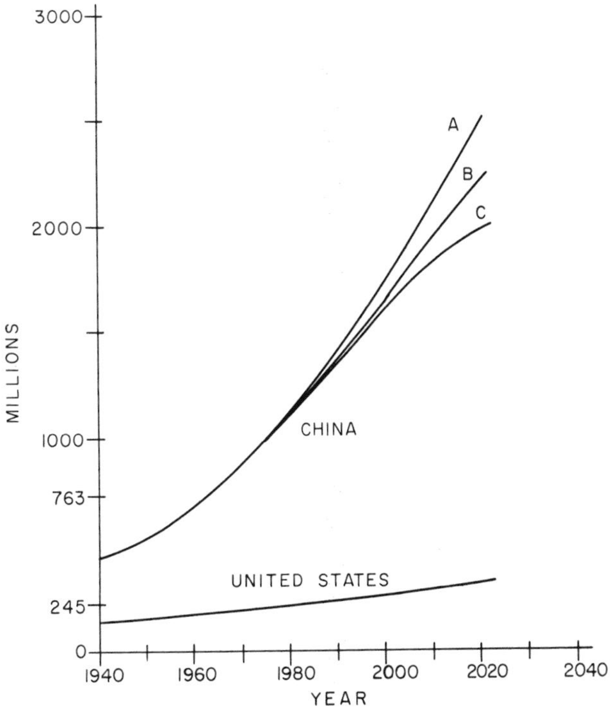

Fig. 6.2. Population estimates, U.S., PRC.

world, and China [19, 21]. The general shape of China's curve is very similar to that for Europe up to the present time. If the shape of the ensuing curve is to reproduce that of Europe, curve A is the resulting predicted curve. Curve B is computed for a symmetrical curve. A similar curve for estimated changing values of American population coefficient would be considerably less than that of China, and it would be positioned to the left of China's curve. Notice also the effect of the Chinese population on the curve for the entire world.

Then, Fig. 6.2 presents estimates of Sino and American populations for the period 1940 to 2040. Here, curves A and B are population estimates computed according to the previous discussion from the data given by curves A and B in Fig. 6.1; curve C represents an average of the other two curves. By the year 2000, China's population should be about 1.7 billion. Interesting biological constraints of population are discussed by Markert [22].

Except for the United States, Europe, and Australia, most of the world will be experiencing a smaller but comparable population explosion to that of China. For example, the populations of North America and Latin America are nearly balanced today, but by the year 2000 there will be two Latin Americans for every North American.

Now that a demographic analysis has been established for the case study, Sino industrial growth, economic resurgence, and potential power are studied and related to the bilinear model in a simple manner. First consider electric power generation.

6.3.2 Power Generation

Electrical power generation forms the arteries of a nation's industrial complex and has become an accepted economic barometer. In 1967, China had planned to generate about 222 million MW-hr of energy. Though there is doubt that China reached this output in 1967, it represents approximately the annual U.S. production during World War II and the Soviet production of 1958 (after the launching of Sputnik I). The U.S. planned to generate about 1240 million MW-hr of electrical energy in 1967, but its rate of annual increase was only about seven percent. Meanwhile, Chinese electrical production was increasing nearly 25 percent annually.

Power generation, as presented in Table 6.2, lends an excellent clue to the rising Sino prosperity. During the 1950s, electrical power generation doubled about every three years. Then, after a brief recession, power production continued to increase nearly exponentially according to a simple first-order differential equation like that of (1.3).

As a nation matures, its economic time constant or its bilinear control changes, just as the population coefficient changes. In other words, saturation eventually constrains the growth in a gradual manner. While many western nations may be approaching saturation both in economics and population, others, such as China, have only recently phased into the period of expansion.

Does China have the necessary resources to develop such a modern network of power generating stations? Certainly she has the people, which is the most basic ingredient, and these people are being trained in

TABLE 6.2

Electrical Energy[a]

Year	Electrical energy (BW-hr)[b]	Generating capacity (MW)
1950	4500	1860
1952	7200	1960
1954	11,200	2600
1956	15,200	3600
1957	19,300	4600
1958	27,500	—
1959	41,500	—
1962	30,000	—
1964	36,000	13,000
1965	(60,000)[c]	33,000
1967	(222,000)[c]	—

[a] See Chao [23].
[b] BW-hr denotes begawatt-hours.
[c] Represents gross estimates.

the trades and sciences. Since the entire Sino industry will expand along with the power generation, and since only a very small portion of the output goes to personal consumption, it is likely that there will be funds to support continued expansion. Furthermore, China has the necessary fuel and water resources. China's potential hydroelectric capability alone is estimated at 300,000 MW [24], and in 1967, China planned to produce nearly half her electricity from hydrogenerating stations.

China's resources in coal, oil, and uranium further enhance her future in the production of energy. Electrical power production from nuclear plants may well evolve as a by-product of China's nuclear weapons program, just as it has in the United States.

Besides power generation, steel production is an excellent index of industrial performance.

6.3.3 Steel Production

PRC steel production in 1966, an estimated 15 million tons, was almost 60 times that of 1950. Figure 6.3 shows that China's steel

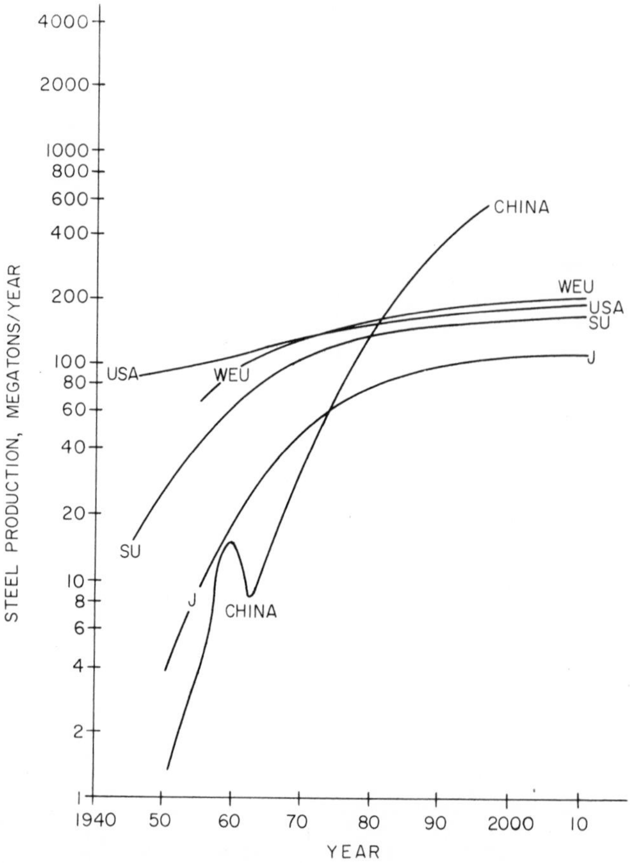

Fig. 6.3. Estimated steel production of several major powers, where J is Japan, SU is Soviet Union, and WEU is Western European Union.

production doubled about every three years (for a steel time constant of about four years) from 1950 to 1958. This growth was very similar to that experienced by the electrical power industry and was generally characteristic of her investment policy which yielded constant control in the bilinear equation [similar to (1.3)]. After launching the Great Leap, the policy changed, and production reportedly made tremendous strides (though somewhat chaotic) in the latter 1950s. After 1960, however, steel production decreased rapidly as a result of overexpansion and ineffective local furnaces.

Though the Chinese steel industry is not as automated as the U.S. industry, improvement is evidenced by many new innovations and increasing labor production figures. From 1952 to 1960, pig iron production increased from 190 tons to 450 tons per man-year of labor. The ability of China to support an expanding industrial program is indicated by her growing steel production.

Steel production of several world leaders is compared in Fig. 6.3. As for electrical energy, increasing steel production for the PRC appears to be exponential, with a time constant of about five years (doubling time of 3.5 years).

Predictions of future steel production are also presented in Fig. 6.3. The data follow that presented by Fucks [21] with a slight correction made for more recent Sino data. Again, the predictions may be computed from a simple first-order bilinear model. Since wages represent only a small portion of the gross value added to the Chinese steel industry, the Communist regime can continue to expand her steel production and still realize a profit for other investments. For example, while added annual investments in wages in the United States and the Soviet Union accounted for 65 and 53 percent, respectively, of the total investment, wages only accounted for six to ten percent of the added value to China's steel industry from 1952 to 1956. This adds a catalyst by parametric control to her economy. Evidently China also has sufficient iron ore and coal mining to support her steel production, for she has been exporting large quantities of both.

6.3.4 Coal and Petroleum Production

As would be expected, coal mining is increasing along with steel production and energy production (see Table 6.3). Shansi province probably has the richest coal reserve in China.

In production of petroleum products, the PRC has become basically self-sufficient, and gasoline prices decreased nearly 19 percent in 1965 [26]. In a Shanghai plant, China claims to have one of the world's most modern oil-production process, fermentation dewaxing. In this process the paraffin in crude oil is removed by means of bacteria. The high-quality oil produced here is used in precision instruments at low

TABLE 6.3

FUEL PRODUCTION[a]

Year	Coal (megatons)	Crude oil (megatons)
1949	32	0.12
1950	43	0.20
1955	98	0.44
1957	270	0.5
1959	350	0.7
1961	—	5.8
1964	240	9.4
1965	300	11
1966	—	16

[a] See Chao [23] and Prybyla [25].

temperatures or at high altitudes. Also, high-protein yeast for the food and pharmaceutical industries is a by-product of the fermentation process [27].

China's growth in petroleum also can be modeled roughly by a single exponential solution with a time constant somewhat smaller (faster rise) than her four- to five-year economic time constant experienced in electrical energy production and steel production. Meanwhile, the U.S. time constant in petroleum output of about 37 years indicates that petroleum in the United States may be relatively saturated.

China's production of petroleum products in 1965 was probably only a few percent of U.S. petroleum production, but most of the U.S. production goes to personal consumption. Certainly China will have to continue her rapid expansion in petroleum if she is to become a world industrial power. There is little doubt that China has the necessary resources for this expansion.

6.3.5 Economic Performance Indices

Electrical energy production, steel production, total value of industrial production, and gross national product are all common indices of economic performance. Obviously, it is the objective of the PRC to optimize some such index within the allowable set of constraints.

Gross national product (GNP) refers to the market value of goods and services produced by a nation's economy. Electrical energy and steel production have been analyzed in detail in the preceding sections, and the approximate total value of Sino industrial production in terms of equivalent U.S. dollars is presented in Table 6.4.

TABLE 6.4

SINO PRODUCTION VALUE, BILLIONS OF U.S. DOLLARS[a]

	1952	1954	1957	1960	1961	1962	1964
Modern industry	9.4	14	24	40	17	31	40

[a] See the literature [23, 28–30].

At the bottom of her recession in 1963, the Chinese GNP was estimated to be about the equivalent of 70 billion U.S. dollars. This was roughly the value of America's GNP in the mid-1930s. Today, China's GNP is probably well over 100 billion dollars. Chinese resurgence based on GNP is definite, but it is slower than that based on steel and energy production.

Economic power indices are discussed by Fucks [21]. Here, GNP is considered unreliable due to differences between different social systems in the price itself, in the set of goods produced, and in the quantity of goods produced. The total value of production can be discarded for similar reasons. Good indices of performance should be very basic physical quantities which have basic and steady value. Energy and population are good candidates of this category. Though steel may some day be substituted by other materials, it is closely correlated with basic material production. Hence steel production is a good economic meter. Fucks tries various functions of population, energy production, and steel production that seem to correspond best to the power hierarchy of leading nations from 1960 to 1963. He recommends the most accurate index of those studied to be

$$M = (M_1 + M_2),$$

where $M_1 = k_{1p}{}^{1/3}s$ and $M_2 = k_{2p}{}^{1/3}e$. Here p is population; s is steel production; e is electrical energy production; and k_1, k_2 are constants such that $M_1 = M_2 = 100$ for the United States in 1960. This index makes PRC resurgence appear somewhat faster (relative to other countries) than that depicted by steel production in Fig. 6.3, but the growth is similar.

The reader with a particular interest in socioeconomic studies of the PRC should see the literature [17–35].

6.4 CONCLUSIONS

A cursory broad-brush analysis of China has been presented as a somewhat philosophical example. Nevertheless, it is apparent that parametric control and a variable structure such as that suggested by the bilinear model are significant in socioeconomics. In the discussion of population, energy production, and steel production, the concept of a piecewise linear constant-coefficient model or a bilinear system with piecewise constant control policies is found to be useful. The latter does provide a more intuitive description of overall behavior.

Periods of different politicosocioeconomic policies which alter the bilinear control or parameter value may be divided into the following characteristic periods: (a) rapid but controlled economic growth, (b) overexpansion by "Great Leap," (c) recession, (d) controlled economic growth [similar to (a)], (e) recession due to cultural revolution, and (f) controlled economic growth broadly based on "grass roots" of society.

Detailed modeling and identification of parameters would be satisfying at this point. Unfortunately, the required data are not available. Still, a basis and philosophy have been established for more detailed study. This type of base must be established initially for any such socioeconomic study.

More detailed models are presented for urban dynamics by Forrester [12] and for the agricultural sector of Nigeria by Manetsch *et al.* [16]. In urban dynamics, parametric multipliers, which take the form of such variables as public expenditures, housing availability, tax rate, and jobs [12], are the driving forces of the system.

It is hoped that the presentation made in this chapter, while by no means complete, will provide a base and motivation for a very deserving area of study.

Exercises

6.1 Why is the bilinear model more effective than the linear model for many socioeconomic processes?

6.2 Estimate China's economic time constant based on a first-order linear model of steel production during "the Great Leap." Why is linear system theory appropriate for such a nonlinear or bilinear system?

6.3 Explain how optimal control theory could be introduced for improving economic performance and relate this to "the Great Leap."

6.4 Explain the manner in which tracer experiments such as those studied in Chapter V for physiological modeling might be applied to socioeconomics.

6.5 Patten [36] uses the following compartmental model to describe the above-ground live-plant biomass x_1 for grass land:

$$dx_1/dt = v - u_1 x_1,$$

where v is a growing force involving inputs from sunlight, temperature, and moisture; u_1 is the sum of loss rate coefficients due to harvest rates to consumers, u_{1a}, and for transfer to belowground biomass, u_{1b}, and to standing dead biomass, u_{1c}. Assume v is constant,

$$\begin{aligned} u_{1a} &= 0.04 && \text{on } (0, 1\text{ yr}], \\ u_{1b} &= 0.04 && \text{on } (0, 1/2], \\ &= 0 && \text{on } (1/2, 1] \quad \text{(no transfer after growing season)}, \end{aligned}$$

and

$$\begin{aligned} u_{1c} &= 0.04 && \text{on } (0, 1/2], \\ &= 0.40 && \text{on } (1/2, 1] \quad \text{(majority transferred to standing dead in autumn)}. \end{aligned}$$

Compute $x_1(t)$ on $[0, 1]$ for $x_1(0) = x_{10}$.

6.6 Similar to the process presented in Problem 6.5, x_2, the below-ground plant biomass, may be described by a compartmental model which for one root compartment may be related to the above-ground one-compartmental biomass by

$$dx_2/dt = u_{1b}x_1 - u_2x_2,$$

where $u_2 = u_{2a} + u_{2b}$ and the other terms are described in Problem 6.5. Here u_{2a} is a turnover rate respiration coefficient, and u_{2b} is a microfaunal grazing coefficient. Assume

$$\begin{aligned} u_{2a} &= 0.0014 \quad \text{on} \quad (0, 1/2 \text{ yr}] \\ &= 0 \quad \text{on} \quad (1/2, 1], \\ u_{2b} &= 0.0010 \quad \text{on} \quad (0, 1], \end{aligned}$$

with $x_2(0) = x_{20}$, and the solution $x_1(t)$ from Problem 6.5, compute $x_2(t)$ on [0, 1].

6.7 (a) Approximate the China population coefficient graph in Fig. 6.1 by a rectangular pulse and use this approximation to calculate China's population after 1900.
(b) Select a desired Chinese population for the twenty-first century and give a rectangular pulse to realize such an equilibrium population in the year 2000.

6.8 Analyze the relative merit of five economic performance indices.

6.9 For the population dynamics

$$dx/dt = ux,$$

where $x(0) = 1.0$, find the $u(t)$ which drives the population to $x(2) = \frac{1}{3}$ so as to minimize

$$J = \tfrac{1}{2}\int_0^2 (x^2 + u^2)\, dt.$$

REFERENCES

1. Keynes, J. M., "General Theory of Employment, Interest and Money." Harcourt, New York, 1935.
2. Runyan, H. M., Cybernetics of economic systems. *IEEE Trans. Syst. Man Cybernet.* **SMC-1**, 8–17 (1971).

3. Harrod, R. F., An essay in dynamic theory. *J. Royal Soc. Econom.* **49**, 14–33 (1939).
4. Domar, E., Capital expansion, rate of growth, and employment. *Econometrica* **14**, 137–147 (1946).
5. Frisch, R., Propagation problems and impulse problems in dynamic economics. *In* "Economic Essays in Honor of Gustave Cassel." Allen & Unwin, London, 1933.
6. Samuelson, P. A., Interactions between the multiplier and analysis and the principle of acceleration. *Rev. Econom. and Statist.* **21**, 75–78 (1939).
7. Metzler, L. A., The nature and stability of inventory cycles. *Rev. Econom. and Statist.* **23**, 113–129 (1941).
8. Phillips, A. W., Stabilization policy in a closed economy. *J. Royal Soc. Econom.* **64**, 290–323 (1954).
9. Goodwin, R. M., A model of cyclical growth. *In* "AEA Readings in Business Cycles" (R. A. Gordon and L. R. Klein, eds.), pp. 6–22. Irwin, Homewood, Illinois, 1965.
10. Duesenberry, J. S., "Business Cycles and Economic Growth." McGraw-Hill, New York, 1958.
11. Forrester, J. W., "Industrial Dynamics." MIT Press, Cambridge, Massachusetts, 1961.
12. Forrester, J. W., "Urban Dynamics." MIT Press, Cambridge, Massachusetts, 1969.
13. Forrester, J. W., "Principles of Systems." Wright-Allen Press, Cambridge, Massachusetts, 1968.
14. Jaeckel, M. T., Forrester's urban dynamics: A sociologist's inductive critique. *IEEE Trans. Syst. Man Cybernet.* **SMC-2**, 200–216 (1972).
15. Gibson, S. E., A Philosophy for urban simulations. *IEEE Trans. Syst. Man Cybernet.* **SMC-2**, 129–138 (1972).
16. Manetsch, T. J., Hayenga, M. C., and Halter, A. N., Simulation of Nigerian development: Northern region model. *IEEE Trans. Syst. Man Cybernet.* **SMC-1**, 31–42 (1971).
17. Mohler, R. R. "China's World Revolution and the Three-Power Equilibrium." To be published.
18. Lee, J. S., The periodic recurrence of internecine wars in China. *China J.* **14**, 114, 1931.
19. Ho, P.-T., "Studies of the Population of China." Harvard Univ. Press, Cambridge, Massachusetts, 1957.
20. Taeuber, I. B., Population: Riddle of past, enigma of future. *In* "Modern China" (A. Feuerwerker, ed.), Chapter 2. Prentice-Hall, Englewood Cliffs, New Jersey, 1965.
21. Fucks, W., "Formeln zur Mach." Deutsche Verlags-Anstalt, Stuttgart, 1965.
22. Markert, C. L., Biological limits on population growth. *BioScience* **16**, 859–862 (1966).
23. Chao, K., "The Rate and Patterns of Industrial Growth in Communist China." Univ. of Michigan Press, Ann Arbor, 1965.

24. Water resources of China (Plans and prospects for development from Japanese). Rep. No. JPRS-32681. U.S. Dept. of Commerce, Washington, D.C., 1965.
25. Prybyla, J. A., Communist China and petroleum. *Mil. Rev.* (1967).
26. China. "Far Eastern Economic Review, 1966 Yearbook," pp. 147–165. Far East Econ., 1966.
27. Fermentation dewaxing. *China Reconstructs*, p. 24, April 1967.
28. Wu, Y.-L., "Economic Development and the Use of Energy Resources in Communist China." Praeger, New York, 1963.
29. Wu, Y.-L., "The Economy of Communist China." Praeger, New York, 1965.
30. An economic profile of mainland China. *General Economic Setting, the Economic Sector*, Vol. I. (A Congressional Rep.) U.S. Gov. Printing Office, Div. of Public Documents, Washington, D.C., 1967.
31. Uchida, G., Technology in China. *Sci. Amer.* **215**, pp. 37–45 (1966).
32. China. "Far Eastern Economic Review, 1967 Yearbook," pp. 121–138.
33. China today (special June issue). *Bull. At. Sci.* **22**, (1966).
34. Stucki, L., "Behind the Great Wall." Praeger, New York, 1965.
35. Cheng, C.-Y., Scientific and engineering education in communist China. Rep. of the Nat. Sci. Foundation, Washington, D.C., 1965.
36. Patten, B. C., A state space model for grassland. "Systems Analysis and Simulation in Ecology," Vol. III. Academic Press, New York, to be published.

AUTHOR INDEX

Numbers in parentheses are reference numbers and indicate that an author's work is referred to although his name is not cited in the text. Numbers in italics show the page on which the complete reference is listed.

SUBJECT INDEX